The Climate Mandate

The Climate Mandate

Walter Orr Roberts

Aspen Institute for Humanistic Studies

Henry Lansford

National Center for Atmospheric Research

W. H. Freeman and Company

San Francisco

Cover photograph on softcover edition: High plains
of northeastern Colorado. (Henry Lansford, National
Center for Atmospheric Research.)

Frontispiece: Summer thunderstorm building over
the plains east of Boulder, Colorado. (Charles Semmer,
National Center for Atmospheric Research.)

Sponsoring Editor: Gunder Hefta
Project Editor: Patricia Brewer
Copy Editor: Susan Weisberg
Designer: Perry Smith
Production Coordinator: Linda Jupiter
Illustration Coordinator: Batyah Janowski
Artist: T. Keenan
Compositor: Allservice Phototypesetting Company
Printer and Binder: The Maple-Vail Book Manufacturing Group

Library of Congress Cataloging in Publication Data

Roberts, Walter Orr.
   The climate mandate.

   Bibliography: p.
   Includes index.
   1. Climatic changes—Social aspects. 2. Crops
and climate. 3. Demography. I. Lansford, Henry,
1929–   joint author. II. Title.
QC981.8.C5R6   301.31   78-25677
ISBN 0-7167-1054-4
ISBN 0-7167-1055-2 pbk.

9 8 7 6 5 4 3 2 1

Contents

Preface

THIS BOOK IS A JOINT EFFORT by a professional scientist and a professional journalist. We have tried to bring to it the best of both professions, to have authority without pedantry and clarity without triviality.

We wrote the book because we believe it is critically important for as many people as possible to understand and accept the fact that the quality and character of human life over the coming decades will be determined largely by three closely related influences: climatic variation, food production, and population growth. Although each of these elements is the product of many interactions and they interact with one another in complex ways, we believe that it is possible for an intelligent reader who is not a specialist in any of these fields to understand the nature and extent of their interactions and impacts on the lives of all the people of our planet.

The reader will find a good bit of history, political science, and economics mixed in with our meteorology and climatology because we have tried to transcend traditional boundaries between academic and scientific disciplines. We think that the only way to understand the broad significance of many important new findings in the atmospheric sciences is to consider them in a societal context.

Our goal was to make the book easy to read and understand without making it simplistic or excessively elementary. Scholarly paraphernalia, such as footnotes and numbered references, have been kept to a minimum, but we have supplied bibliographical information for the reader who wants to dig deeper into subjects that we have not examined exhaustively. We have also interviewed many people, and some of our quotations do not have sources listed in the bibliography because they came from interviews rather than the printed page.

We are convinced that the problems considered in this book are of critical importance to the future of humanity. We hope that the book will contribute to an understanding of these problems and thus to humankind's ability to cope with them.

A great many individuals and institutions have made substantial contributions of time and knowledge to this book. We would like to express special appreciation to Stephen Schneider and Michael Glantz, both of the National Center for Atmospheric Research, and to Louis Thompson of Iowa State University. Other specific contributions are acknowledged at appropriate places in the text. However, many other friends and colleagues have contributed in broader but no less significant ways, through discussions at scientific meetings, conversations over coffee, and other kinds of formal and informal interactions. You know who you are; please accept our thanks for your help.

December 1978   WALTER ORR ROBERTS
                                                  HENRY LANSFORD

The Climate Mandate

CHAPTER 1

Harvests and Human Survival

F OR MANY MILLIONS of the earth's four billion people, the single most important event that occurs each year is the harvest—if the harvest is good, they will live another year; if it is bad, many of them will die.

Many of us who read books such as this one have never paid much attention to the harvest. We are members of the world's affluent minority, and our worst food problem is the need to diet and exercise to keep from getting too fat or falling victim to hypertension, heart disease, and other maladies of the overfed. Our highly organized social systems and heavy use of technology to manipulate the natural environment have created the illusion that we are insulated from such elemental problems as whether the wheat, corn, or soybean harvests are good or bad. We have the habit of thinking of the supermarket, not the land, as the source of our daily bread.

But it's not that way everywhere. Hundreds of millions of the world's people practice subsistence farming and depend on each year's harvest to feed the farmers and their families for the next year. If the crops fail, they go hungry. The statistics on world hunger and malnutrition are frightful. An estimated one billion people suffer from malnutrition, and 400 million live on the brink of starvation. Infants and young children are particularly vulnerable—perhaps 10 million children around the world are so badly underfed that their lives are constantly in danger; each year

Harvesting rice in the Upper Solo River Basin of Indonesia, one of many regions of Asia where people depend on a good rice harvest to survive from one year until the next. (United Nations/ Food and Agriculture Organization.)

an estimated one million children die from malnutrition in India alone. Those who survive to maturity often suffer irreparable physical and mental damage because of childhood malnutrition.

Each year our planet needs more food, for a simple reason: Each year there are more people to feed. Today, there are 200,000 more people in the world than there were yesterday; tomorrow there will be 200,000 more. Increasing affluence in developing nations accounts for about one-fifth of the world's current rate of increase in food demand, but the other four-fifths results from population growth. Between 1930 and 1976, the earth's population doubled from two billion to four billion, with six billion expected by the year 2000. Although the birth rate has dropped off in the United States and some other affluent countries, it continues to increase in regions where malnutrition is already a tragic problem—the poor countries (or developing nations, in the parlance of international development) of Asia, Africa, and South America.

The idea that population can outrun food supplies is not new. In 1798, Thomas Malthus, an English clergyman, wrote "An Essay on the Principle of Population." He maintained that population, which increases exponentially, doubling in a given period of time, must inevitably outrun available food, which he assumed could increase only by a fixed amount during that same period. Unless population growth was deliberately checked, Malthus maintained, it ultimately would be controlled by starvation, disease, poverty, and other forms of human misery that would result from overpopulation.

The earth's population has continued to grow exponentially, but Malthus's prophecy has yet to be fulfilled, although the trend was in that direction for many years. It has been estimated that less than two million people starved to death in the seventeenth century, but that this figure shot up to 10 million in the eighteenth century and 25 million in the nineteenth. An estimated 12 million deaths from starvation have occurred during the first three-fourths of the twentieth century, and another 500 million people have probably died from malnutrition-related diseases.

Our first reprieve from a Malthusian catastrophe was gained by the opening of new lands for cultivation during the last part of the nineteenth century and the early years of the twentieth. Much of the fertile and productive Great Plains region of North America—sometimes called the breadbasket of the world—came under the plow during that period of expansion.

But even with new land under cultivation, there would have been no way to feed four billion people if agricultural productivity—the amount of grain or other crops yielded by each acre or hectare of farmland—had remained at the levels of the early 1900s. Remarkable increases in productivity have been achieved during the past 25 years or so. In the United States, corn yields have doubled since 1950, and yields of many other crops have also increased spectacularly, mainly as a result of heavy use of nitrogen fertilizer and the introduction of new varieties of grain developed by plant breeders. In the developing nations, the combined package of new high-yield grains and improved agricultural practices usually called the "Green Revolution" has brought great increases in productivity.

The agricultural scientists responsible for the Green Revolution were the first to emphasize that they had simply bought a little more time for humanity to try to solve the problem of runaway population. The dramatic increases in world agricultural productivity of the last few decades clearly cannot continue indefinitely at the same rate. For one thing, most of the good farmland is already under cultivation; what remains is mostly marginal in terms of soil quality, water supply, or other critical factors. Crops can be grown almost anywhere given sufficient water, nourishment, and cultivation, but production costs and the chances of crop failure both rise sharply as the arability—the suitability for agriculture—of the land decreases. The Great Plains of North America, the Ganges-Bramaputra River Plain of India, the rich farmlands of western Europe, and most of the fertile and well-watered land in the world lies in parts of the Northern Hemisphere that are already under cultivation. Most of the arable land that remains is in parts of tropical Africa and South America that will be difficult to utilize as food-producing lands because of problems of politics, transportation, and other factors in addition to cultivation. Ironically, the regions where population growth and food demand are highest are also those where good farmland is scarcest. Over the last few decades, the people of those regions have become increasingly dependent on imported food for survival. With each passing year, the harvest in critical exporting regions such as the North American breadbasket has become more and more important to hungry people all around the world.

For as long as people have cultivated the land to raise food, the most important factor that has determined whether the harvest is good or bad has been the weather, or, in long-range terms, the climate, which is weather writ large—the summation of weather conditions over a period of time. One question that has been debated in the past, and that becomes increasingly critical as more marginally arable lands are brought under cultivation, concerns the role that weather and climate have played in the increases in agricultural productivity of the last quarter century. Some agricultural scientists have maintained that the same advances in agricultural technology that increased productivity also reduced the vulnerability of crops to the vagaries of weather and climate. But there is evidence that this is not so, and that an unprecedented run of good North American growing weather in the 1950s and 1960s was the major factor in the uniformly high yields of U.S. agriculture during that period. If this is so, then bad weather would have the opposite effect, and food production could drop significantly.

Our understanding of the mechanisms involved in short-term fluctuations and long-term changes in climate is inadequate to make any sort of knowledgeable and credible prediction of what the weather and climate will be like over the next years, decades, or centuries. But some atmospheric scientists see a strong probability that the climate of the future could be quite different from that of the recent past. Even if no long-term climatic changes, such as new ice ages or dramatic warming trends, are imminent, the kind of climatic variability that resulted in anomalies such as the U.S. Dust Bowl drought of the 1930s could have a serious impact on future harvests.

Those of us who live in the rich nations take our harvest for granted, as manna provided by the great god Technology. In the United States, for example, agricultural production has risen so much during our lifetimes that the government has paid farmers not to grow crops on their land. With huge surplus stocks of grain in storage, we were protected against the consequences of bad harvests.

But in the early 1970s the surpluses vanished, and for several years world grain reserves were precariously low. Now we are beginning to listen more closely to news of the harvest. The rising food prices and political controversy that followed massive purchases of North American grain by the Soviet Union after that country's poor harvest in 1972 got our attention. It's becoming clear to most of us that the success or failure of the world's harvests is starting to have a direct and increasing impact on the quality and character of our lives. Good and bad harvests probably won't mean the difference between life and death for most of us in North America, Western Europe, and other affluent industrialized regions of the world, but they may mean the difference between prosperity and poverty, social stability and turmoil, or war and peace.

Thus climate imposes a mandate on us—a compelling pressure to deal with the problem of too many people and not enough food as soon as possible and as effectively as possible. If we wait to see what happens, on the theory that population growth and food supplies are converging gradually and predictably, the climate may intervene with a couple of bad years, or even more, in a row, as it did in 1972, when drought and other weather problems caused massive crop failures around the world.

If we try to ignore the climate mandate, the chances are good that hundreds of millions of people will die unnecessarily, and the lives of hundreds of millions of others will be affected in ways that range from trivial annoyances to tragic disruptions.

CHAPTER 2

Weather, Climate, and Human Affairs

AS TECHNOLOGY made us practically invulnerable to assaults from the atmosphere? Are weather and climate insignificant factors in the conduct of human affairs in the second half of the twentieth century? Rather than trying to consider these questions in abstract and general terms, let's begin by looking at some recent manifestations of weather and climate and considering their impact on the people who lived where they occurred.

*   *   *

THE LATE 1960s seemed to be a time of promise for the people of the Sahel, the semiarid steppe that stretches from the west coast of Africa eastward toward Ethiopia, bounded on the north by the Sahara and on the south by the grassland and scattered trees of the region known as the Sudan.

The Sahel is crisscrossed by the boundaries of half a dozen independent nations—Mauritania, Senegal, Mali, Upper Volta, Niger, and Chad—carved out of the old colonial empire of French West Africa. Many of the people of the Sahel are nomadic herders, organized into ancient tribal confederations that bear no relation to the new national boundaries. For many centuries, they have migrated with the seasons to find pastures for their herds.

In the 1960s, foreign aid brought the beginnings of a decent health-care system to the Sahel, and the human population began to rise as the death rate went down. The cattle population was increasing too. Cattle

This malnourished West African child is one of perhaps 10 million children around the world who are so badly underfed that their lives are constantly in danger. (Agency for International Development.)

are more than a source of food for the pastoral tribes of the Sahel—they are a bank account and a visible manifestation of an individual's success and affluence. When Sahelian herders prosper, they traditionally put their profits into more cattle.

The size of Sahelian herds had always been constrained by the limited water supply; although the region is not as dry as the Sahara to the north, water is never plentiful. A tribe could keep only as many cattle as it could supply with water hauled up by hand from shallow wells (Figure 1). But foreign aid changed that, too. The U.S. Agency for International Development (AID) drilled more than 1400 deep wells in the region, tapping a vast subterranean water supply that accumulated many centuries ago. French and British technical assistance groups also drilled deep wells. The wells were equipped with pumps driven by engines, and with this new and generous supply of water the herds of cattle grew very large in the areas around the wells. Soon there were too many cattle for the sparse forage that was now being steadily grazed instead of having a chance to recover between visits by nomadic herds. But six years of reasonably good rainfall in the mid-1960s improved the pasture land as the herds grew larger, so that the imbalance that was developing was not apparent.

In 1968, there was a change in the pattern of summer weather in the

Sahel—the monsoon did not push as far north as usual. The monsoon is the great wind system that normally sweeps the summer rains northward into the Sahel in July and August, carrying moist air up from the tropical Atlantic. Like some other tropical regions in Southeast Asia and the Indian subcontinent, the Sahel depends on the summer monsoon for moisture to water its pastures and crops; very little rain falls during the rest of the year. The weak monsoon of 1968 brought scanty rainfall to the Sahel, far below the levels of the previous six years. The drought was not exceptionally severe compared to some that had struck the area in earlier years of this century, but its ecological impact was devastating. As the overgrazed pastures turned brown and bare, the great herds foraged for food wherever they could find it. The goats devoured the roots from the ground; the cattle ate the foliage from the trees.

Although more rain fell on the Sahel in 1969 than in 1968, rainfall that year was still well below the levels of the mid-1960s. The hope that the drought might be a brief and transient episode vanished as the rains failed again in 1970, and again and again and again for the next three years. By the time the summer rains returned to break the drought in 1974, thousands of square miles of pastures and farmland had become barren wasteland, the great herds of the Sahel had been reduced by starvation to a fraction of their former size, and the worst Sahelian famine of the century had killed more than 100,000 people in spite of a massive international relief effort that saved thousands of others.

*   *   *

EARLY IN AUGUST 1969, an instability appeared in the atmosphere somewhere off the west coast of Africa. It developed into a tropical disturbance—a cluster of clouds and upward-moving air driven by heat from the tropical ocean. Traveling with the trade winds, this weather system drifted westward over the Caribbean Sea. As it passed south of Cuba, it began to grow more intense, and soon the meteorologists at the U.S. National Hurricane Center in Miami gave it a name: Camille.

Several months, later, the center's director, Dr. Robert H. Simpson, would call Camille "the greatest storm of any kind ever to have affected the mainland of the United States" (Figure 2). Another hurricane specialist, Clyde Connor, who headed the hurricane forecasting operation at

```
YR MO DY HR MIN SC TK ZO S ESSA    M C LAT SP LONG SP ORBIT FR  SUN GLINT,
69  8 17 19  57  28  3 60   9      7 2 35N  5  90W  5  2154  5  31N   95W
```

FIGURE 2   A satellite photograph of the great swirling vortex of Hurricane Camille just before its eye crossed the Mississippi Gulf Coast on August 17, 1969. (National Oceanic and Atmospheric Administration.)

New Orleans, described Camille as "the most intense hurricane of record to enter the U.S. mainland" (Sugg and Pardue, 1970).

The eye of Hurricane Camille crossed the Mississippi Gulf Coast at about 10:00 PM on August 17. Just before the storm moved inland, winds at the surface were estimated at close to 200 miles per hour. Driven by the powerful winds, the waters of the Gulf of Mexico surged over the coast in tides that reached heights of 15 to 20 feet between Bay St. Louis and Biloxi, Mississippi. The wind and water battered down everything in their path, leaving a wake of almost total destruction along the heavily developed Mississippi and Louisiana coastline (Figure 3).

As they move inland and lose their supply of heat and moisture from the warm sea, hurricanes usually decrease rapidly in intensity. Camille

FIGURE 3   The force of Hurricane Camille completely demolished many homes along the Gulf Coast and piled up others. (National Oceanic and Atmospheric Administration.)

weakened as it moved northeast across Mississippi, Tennessee, and Kentucky. But then the storm revived unexpectedly, producing torrential rains and disastrous floods in the upper James River basin of Virginia. In the late hours of August 19, the rains descended—more than 2 feet of rain dropped on some locations in about 8 hours, with a record-breaking 31 inches in the vicinity of Lynchburg and Scottsville, Virginia. A U.S. Weather Bureau hydrologist characterized this storm as one of nature's rare events: "Rainfall of this magnitude occurs, on the average, only once in more than 1000 years," he said. The Virginia flood took 107 lives, with 55 more people missing and assumed dead. Many died as they slept, their homes swept away by rampaging streams and massive landslides (Thompson, 1969).

Altogether, Hurricane Camille killed more than 250 people and caused record losses of nearly 1.5 billion dollars.

*　　*　　*

DURING THE SUMMER OF 1972, a period of very hot, dry weather in the Soviet Union devastated crops in the region around Moscow, adding to agricultural problems that had begun with a lack of winter snow cover in the wheat-growing regions of the Ukraine. In India, the monsoon came one to two weeks late and retreated early, leaving the country's wheat and other crops short of badly needed moisture. The drought in the Sahel continued, and there were serious droughts in other parts of Africa and in South America and floods in the midwestern United States. For world agriculture, it was the most widespread and serious episode of bad growing weather in recent times.

Total world food production dropped by more than 2 percent in 1972, the first decrease since World War II. The drop in food production was sharpest in developing countries such as India and Pakistan, and its effects were most serious there—famine spread in many parts of those nations, as well as in other Asian and African countries. There were also impacts in the developed nations, mainly on trade patterns and prices. The Soviet Union made up part of its food deficit by purchasing 28 million tons of grain from the United States and Canada, reducing grain reserves, raising the prices of meat and bread in U.S. supermarkets, and stirring up a sizable political controversy over this country's agricultural and export policies.

*　　*　　*

DURING 12 HOURS on April 3 and 4, 1974, nearly 150 tornadoes lashed 14 states in the southern and midwestern United States (Figure 4), breaking the previous record of 69 tornadoes during a 24-hour period. According

FIGURE 4   On April 3 and 4, 1974, nearly 150 tornadoes like this vicious midwestern twister roared across 14 southern and midwestern states. (National Oceanic and Atmospheric Administration.)

to Allen Pearson, director of the National Severe Storms Forecast Center in Kansas City, more than 50 of these twisters were "super tornadoes" of exceptional strength and potential for destruction.

One of the most severe tornadoes in the outbreak struck Xenia, Ohio. It cut a swath of destruction 16 miles long and half a mile wide, killed 32 people, and damaged or destroyed 2400 homes and most of Xenia's business district. According to Pearson, all the evidence suggested that the sheer magnitude of the 3–4 April 1974 outbreak had never before happened in this century. David Ludlum, a student of the history of U.S. weather and climate, says that there has been nothing to compare with this tornado outbreak since February 1884, when at least 60 tornadoes ripped over the southeastern United States during a 14-hour period, killing an estimated 800 people and injuring 2500.

More than 300 people died in the tornado outbreak of April 3 and 4, 1974.

*   *   *

THE YEAR 1974 was not a good one for the farmers of the central United States, a fertile region that supplies a large fraction of the world's food. Early in the year, floods in the Midwest delayed spring planting. While the fields of Iowa were still soggy from the floods, drought was developing over the high plains to the west, and Denver had the driest May in its history. The shortage of rain, combined with very hot weather, wiped out the wheat crop in some parts of the southwestern plains and seriously damaged midwestern corn.

On the first day of autumn, September 23, a sudden cold snap brought the earliest freezing temperatures on record to Detroit and Dayton. In Indiana, the freeze brought an early and abrupt end to the tomato season and severely damaged corn and soybeans that had been planted late because of the spring floods. Corn and soybeans in Iowa, Nebraska, and Wisconsin also suffered serious damage, and the early freeze shortened the growing season in central Illinois to 151 days, compared with an average of 189. The bad weather reduced the U.S. corn crop from a predicted 6.7 billion bushels to around 5 billion and the wheat crop from 2.2 billion to less than 1.8 billion. Other crops also fell short of expectations. United States grain reserves, which had been reduced by the Soviet purchases in 1972 and 1973, fell to their lowest level in 20 years.

*   *   *

IN JANUARY 1975, the Great Plains of the north central United States and south central Canada had one of the worst blizzards in the history of the region. It lasted three days, from Friday, January 10, through Sunday. In

Minnesota, the highway department gave up and pulled in its snowplows after snowdrifts started to build up to heights of 20 feet, and for several hours snowmobiles were the only vehicles that could move. In South Dakota, the combination of sustained high winds and extreme cold was the worst recorded in this century. The wind chill factor produced the equivalent of temperatures of −50 to −60 degrees Celsius (−60 to −80 degrees Fahrenheit). In the five-state area hit by the blizzard—the Dakotas, Iowa, Nebraska, and Minnesota—more than 55,000 head of cattle died, frozen by the subzero cold, buried by drifting snow, or suffocated to death when their nostrils were blocked by frozen moisture. Eighty people in the five states died as a result of the blizzard, either from direct causes, such as exposure, or from heart attacks resulting from their efforts to fight the snow and wind. Meteorologists as well as old timers said there had been nothing like this storm since the Great Blizzard of 1888 (Graff and Strub, 1975).

*   *   *

THE WINTER OF 1974–1975 was mild in England, and jonquils bloomed in February. But snow fell on London in June for the first time in this century. In July and August of 1975, Great Britain and Europe had the hottest weather in many years. London bobbies went tieless for the first time in history. Consumption of English beer, which by tradition is served at room temperature, went down as sales of cold lager rose by 60 percent, and Dalek Death Rays Ice Lollies—an iced confection on a stick—sold at the rate of two million a week. Mollie Painter-Downes, the *New Yorker* magazine's London correspondent, described conditions there as "what sunburned Londoners call the splendid summer and water-board officials call the calamitous drought."

Fires raged through parched forests in Germany, Italy, and other parts of Europe. While grapes in the wine-growing regions of France and Germany grew rich and heavy on the vines, other crops suffered from the hot, dry weather. Grain, potatoes, and other food crops were damaged, and milk production dropped because hayfields were drying up. Parts of Brittany and Normandy were declared agricultural disaster areas by the French Government.

*   *   *

BY THE FALL OF 1975, reports had started to come from the Soviet Union about bad weather and a poor harvest there, and they were reinforced by Soviet efforts to buy grain abroad. Early in October, the Soviet Minister of Agriculture, Dimitrii Polyanskii, admitted in a public speech that "unusual weather conditions" had affected this year's Soviet agricultural production. There were rumors that the 1975 harvest might fall even lower than the disastrous levels of 1972.

In the United States, dock workers threatened to boycott grain shipments to the Soviet Union unless they had some assurance that the sales would not raise consumer prices in the United States. An embargo on further sales of U.S. grain to the Soviet Union, imposed in August by President Ford, was lifted in October when a long-term agreement was completed between the two countries. This pact called for the United States to sell up to 7 million metric tons of corn and wheat to the Soviet Union by September 30, 1976, and for the Soviets to buy a minimum of 6 million metric tons annually for five years.

Oren Lee Staley, president of the National Farmers Organization, called the agreement "outrageous and illegal interference with farm exports, not just for this year, but for the next five." Secretary of Agriculture Earl Butz disagreed. "Unexpected Russian purchases have been one of the great destabilizing forces in the world grain markets," he said. "We're just trying to crank in Russia as a regular customer like the Japanese."

*   *   *

DURING THE WINTER OF 1975–1976, the weather in Britain and western Europe continued warm and dry. In May, the British Meteorological Office announced that the period from May 1975 to April 1976 had been the driest year on record. June rainfall in England and Wales was less than 40 percent of the long-term average.

On a Saturday in late June, thermometers in central London hit a record 35 degrees Celsius (95 degrees Fahrenheit), and tempers rose with the temperature. A police spokesman said: "We've had three times as many punch-ups in bars as we usually do." London's ambulance service reported one of their busiest days dealing with domestic squabbles, bar brawls, and street fights, and one shocked ambulance service officer declared that "The sun has turned us into a different race."

As July came, the heat continued. Paris bus drivers found their vehicles so stifling that many of them walked off the job. Barges on the Rhine in West Germany were carrying half-loads because the water level had dropped so low. Industrial cutbacks, in the form of three-day work weeks and layoffs of factory workers, were instituted in South Wales, northern Italy, and other regions where water supplies were critically low.

Production of potatoes, sugar beets, and other crops was down 20 percent in English farming areas. The French grain harvest was expected to be off by 25 percent or more, and French cattle raisers took their animals to the slaughterhouse as the pastures gave out.

Cold lager continued to outsell the traditional English pint o' bitter, and British wine growers were in good spirits as the hot, dry weather promised a bumper crop of grapes.

*   *   *

EARLY ON THE EVENING of July 31, 1976, it started raining hard in Big Thompson Canyon, a narrow, scenic gorge that winds down toward the plains from the town of Estes Park, at the eastern edge of the Colorado Rockies. It was an incredible downpour; by 11:00 PM, more than 10 inches of rain fell just below Glen Comfort, one of a handful of little communities strung out along the canyon highway, U.S. 34. The average annual precipitation in this area is less than 16 inches (Henz and Sheetz, 1976).

On that summer evening, there probably were more than 2000 people in the Big Thompson Canyon—fishermen, campers, families returning from a weekend drive in the mountains, and people living in summer cabins and permanent homes along the banks of the pretty mountain river called the Big Thompson.

As darkness fell, drivers on the canyon highway started to have trouble seeing the road ahead through the driving rain. Then some of them realized that water was running across the road and rising fast. Sections of the pavement began to crumble off into the river, and cars were swept away. Bridges collapsed, riverside campgrounds vanished into the night, and buildings dropped away into the black water. By midnight, the river had taken the canyon (Figure 5).

On Sunday morning, it was clear that the Big Thompson flash flood had been a terrible natural disaster, but the extent of the death and destruction emerged slowly. The front-page headline in Monday's *Denver Post* said: "60 die; Hundreds Stranded." Tuesday's headline was: "88 known Dead in Flood; Final Survivors Rescued." Over the next two months, 139 bodies were found in the mud and wreckage in Big Thompson Canyon, and there were more than a dozen names still on the list of missing persons thought to have been in the canyon on the night of July 31.

*   *   *

These vignettes show some notable weather and climate events of the past decade and their impacts—ranging from the trivial to the tragic—on the quality and character of human life. They raise several provocative questions. The first one that may occur to many readers is perhaps the simplest to answer: What is the difference between weather and climate?

FIGURE 5   The flash flood that roared down Colorado's Big Thompson Canyon on the night of July 31, 1976, chopped the highway out from under many cars and carried away homes and campgrounds, killing at least 139 people. (*The Denver Post.*)

One authority, the British climatologist Hubert H. Lamb (1972), defines the two terms like this:

Weather is taken to mean the totality of atmospheric conditions at any particular place and time—the instantaneous state of the atmosphere and especially those elements of it which directly affect living things.

Climate is the sum total of the weather experienced at a place in the course of the year and over the years. It comprises not only those conditions that can obviously be described as "near average" or "normal" but also the extremes and all the variations.

From the climatologist's point of view, these are satisfactory working definitions. But as we consider the impact of weather and climate on human affairs, it might be wise to extend Lamb's definitions a little. Instead of limiting weather to the atmospheric conditions at a single time and place, let's broaden the definition to include the total behavior of what the meteorologists call a weather system. These systems range in scale from a single thunderstorm or tornado with a lifetime measured in minutes or hours to the great traveling systems known as cyclones, which rotate around centers of low pressure, cover hundreds of square miles, and maintain their identity for days or even weeks. These cyclonic systems, moving across the temperate zones, bring us the familiar patterns of alternating fair and stormy weather over North America, western Europe, and other northern and southern temperate-zone regions.

Likewise, we should not limit our definition of climate too strictly. Rather than thinking in terms of yearly averages and extremes, we should consider seasons as well. A cold winter, a wet spring, and a dry summer are all climate for our present purposes.

By this definition, we would classify as weather Hurricane Camille, the April 1974 tornado outbreak, the upper Great Plains blizzard of 1975, the snow that fell on London in June 1975, and the storm that caused the Big Thompson Flood. By contrast, the Sahelian drought; the worldwide droughts and other anomalies of 1972; the 1974 sequence of flood, drought, and early frost in the United States; and the hot, dry European summers of 1975 and 1976 would be categorized as climate rather than weather. Although severe weather events often result in sudden deaths and violent destruction of property, the impacts of climatic anomalies such as droughts may be more disastrous, if less dramatic, in the long run.

A second question—Is climate changing for the worse?—is not answered so easily. It isn't difficult to make a plausible, if facile, case for worsening weather and climate on the basis of a few dramatic pieces of evidence. After all, Hurricane Camille was, on good authority, the greatest storm ever to have affected the mainland of the United States, and its aftermath, the Virginia cloudburst, produced quantities of rain that occur only once in 1000 years. A tornado outbreak of the magnitude of the one that occurred in 1974, again on good authority, never before happened in this century. Denver had the driest May in its history in 1974, and Detroit and Dayton had the earliest freezing temperatures on record. The 1975 Great Plains blizzard was the worst since the Great Blizzard of '88. And the weather in London had never before, until the summer of 1975, been hot enough for the bobbies to shed their neckties.

The question of whether the climate is in fact changing back in the direction of harsher conditions that existed in previous centuries has been debated extensively and often acrimoniously by atmospheric scientists in recent years. Later in this book we will examine some evidence on both sides of this debate; however, now let's consider another question raised by some of the examples that we cited earlier. If the climate is not clearly and demonstrably worsening, why have so many records, in terms of loss of life, damage to property, and disruption of human activities and institutions, been broken by recent weather and climate events?

There is a straightforward, if somewhat incomplete and over-simplified, answer to this question. The impact of weather and climate on people has increased because the number of people has increased. More people erect more buildings, eat more food, and farm more land, and thus become more vulnerable to the impact of each tornado, hurricane, or drought.

But it's more complicated than that. As the number of people has increased, their artifacts and activities have tended to become more highly concentrated in certain areas. Social, political, economic, and other institutions have become more complex, and linkages between nations and regions have become tighter, so that things that happen in one place increasingly affect many others. A decision by U.S. foreign aid administrators to try to help the people of the Sahel by drilling deep wells became one of a set of converging factors that ultimately increased the impact of the problem—a shortage of water—that it set out to solve. A bad winter and a dry summer in two food-growing regions of the Soviet Union resulted in a grain purchase that upset family food budgets and caused political controversy in the United States.

A fourth question is equally difficult to answer. In a time when science and technology seem to have provided human institutions with many potentially effective new tools to reduce the impact of natural hazards on human affairs, why are those tools often used so ineffectually? Why, in fact, are they sometimes used in ways that ultimately increase the impact of the hazard instead of reducing it?

Two scientists who have devoted a great deal of attention to the impact of natural hazards on human activities and institutions, Gilbert F. White and J. Eugene Haas of the Institute of Behavioral Science at the University of Colorado, sum it up this way:

Research today concentrates largely on technologically oriented solutions to problems of natural hazards, instead of focusing equally on the social, economic and political factors which lead to nonadoption of technological findings, or which indicate that proposed steps would not work or would only tend to perpetuate and increase the problem. In short, the all-important social, economic and political "people" factors involved in hazards reductions have been largely ignored. They need to be examined in harmony with physical and technical factors. It is not a question of more technology or less technology, but of technology in balance.

These last three questions—what is happening to the climate; what are the impacts of climatic fluctuation on the quality and character of human life; and how scientific, technological, and humanistic approaches can be integrated and balanced to reduce the adverse impacts of climate—are what this book is about.

During many periods in the past, glaciers like this one on the slopes of Switzerland's Matterhorn crept down into the valleys and over much of the Northern Hemisphere. (Swiss National Tourist Office.)

CHAPTER 3

Climates of the Past

OULDER, COLORADO, where we—the authors of this book—both live and work, is located at the foot of the eastern slope of the Rocky Mountains. To the east, the semiarid high plains stretch off toward a far horizon. To the west, the Front Range of the Rockies rises sharply toward the 13,000-foot peaks along the Continental Divide.

About 70 miles due south of Boulder, in a high mountain basin northwest of Pike's Peak, stands the little town of Florissant. The surrounding countryside is beautiful in a rather austere way, with scattered stands of ponderosa pine growing on rocky hills that surround grassy meadows. The climate in this valley is semiarid, with an average annual precipitation of about 18 inches, and the winters are harsh and cold.

Florissant stands on an ancient lake bed composed of shale deposits built up of ash spewed out by prehistoric volcanoes that stood near the lake. As it settled to the lake bottom over many centuries, the volcanic ash carried with it leaves, insects, and other small life forms, preserving a remarkable fossil record of what life was like at Florissant 30 to 40 million years ago.

The Florissant shale beds are known to paleontologists around the world for their fossil insects, but a variety of plant fossils have also been found at Florissant. One of the most impressive to many visitors is a huge petrified tree stump, more than 20 feet in diameter at its base and 14 feet tall. It is all that remains of a giant sequoia, similar to the coast redwoods that grow in present-day California. Another fossil tree found at Florissant is a cypress whose nearest living relative, like the coast redwoods, grows only along the Pacific Coast. Other trees and shrubs that once grew at Florissant are species that are found today in moist, subtropical regions of southern Texas and northern Mexico.

Today, the winters are cold at Florissant, and the summers are dry. Redwoods, cypresses, and other species that like warmth and moisture could not survive there. Scientists who have studied the succession of prehistoric plant and animal communities at Florissant, using the evidence preserved in the shale of the fossil beds, see clear indications that, 35 million years ago, the climate at Florissant was moist and warm, similar to today's climate in subtropical northeastern Mexico. But over the period when the shale beds were building up, climatic conditions grew increasingly dry and temperate, shifting toward the conditions that prevail at Florissant today. Thus the petrified redwood and cypress stumps stand as monuments to a sizable long-term change in climate.

*　　*　　*

EIGHTEEN MILES WEST of Boulder and a mile higher, just below the jagged crest of the Continental Divide, the eastern slope of the Front Range drops off into a series of steep, bowl-shaped valleys that open to the east and are separated by sharp east-west ridges. From the air or on a relief map, these valleys look something like the imprint of a giant primeval thumb pressed deep into the rock again and again.

Most of these valleys still contain a clue to their actual process of formation—a patch of ice and snow that stands year-round at the head of the valley. Although these ice bodies bear little resemblance to the huge rivers of ice that are found in the Alps, Alaska, and other regions that are famous for their glaciers, some of them are, in fact, small glaciers. Arapaho Glacier is the source of part of Boulder's water supply, and it is owned by the city. St. Mary's Glacier, west of Denver, is a local landmark of long standing. Visitors to Rocky Mountain National Park can view a quartet of named glaciers—Taylor, Andrews, Tyndall, and Sprague— from the terrace of the Moraine Park Visitor Center.

These little glaciers of the Front Range are fragmentary remnants, or perhaps small-scale recurrences, of huge behemoths of ice that thrust downward from the Continental Divide during glacial episodes of the last million years or so, at the same time that a sheet of Arctic ice was spreading southward, reaching as far as the present sites of New York City and Berlin. It was the huge mass of those giant valley glaciers that crushed and carved the raw granite of the Front Range more than 20,000 years ago, forming the great cirques—glacial valleys—that testify to a time of climate very different from our own: the last period of extensive large-scale glaciation, which ended about 10,000 years ago.

*　　*　　*

THE GRASSLANDS and wheat fields that roll eastward from Boulder are part of the region that is known to most people today as the Great Plains. But in the report of his expedition of western exploration in 1819–1820, Major Stephen H. Long called these high plains the Great American Desert. Later explorers concurred. Lieutenant G. K. Warren, who led a government expedition in the later 1850s, said that the 97th meridian, which runs approximately through Wichita, Kansas, and Lincoln, Nebraska, was the western limit of agriculture, separated by a "desert space" from the arable lands of the Pacific Coast. Major John Wesley Powell, the explorer and self-taught scientist who became the first director of the U.S. Geological Survey, also held this view. In his "Report on the Lands of the Arid Region of the United States," issued in 1875, Powell warned that there was not enough rainfall to grow crops without irrigation west of the 100th meridian.

These conservative views ran counter to the optimistic mood of the post–Civil-War period in the United States. The railroads were pushing west, and there was a strong feeling abroad that the nation's "manifest destiny" was to expand westward. During the 1880s, the erratic rainfall in the high plains was a little more generous than it had been in past years, resulting in a multitude of theories that the climate was growing more favorable for the farmer. The rain belt was moving westward, said the promoters for the railroads. "Rain follows the plow" was the slogan of those who professed to believe that breaking ground for agriculture somehow altered the precipitation processes and brought more rainfall. Homesteaders flocked to the plains, and for two or three years many of them made good crops using the conventional farming methods that had worked back East.

But 1889 was a dry year, and it was followed by more dry years, culminating in a severe drought in 1893–1894. Many homesteaders were wiped out, and thriving little communities that had sprung up as trading centers for the farmers became ghost towns. Then the drought ended, and a new wave of homesteaders came to the plains in the years between 1900 and 1910. Another drought came, but its effects were partly offset by the great demand and high prices for wheat that came with World War I. After the war, larger and larger tracts of grassland were plowed up for big mechanized wheat-farming operations, financed by eastern capital and sometimes even by European investors. It began to look as if Powell and the other pessimists had been wrong—how could this land be unsuitable for agriculture when so much of it was being farmed with such great success?

Then came the granddaddy of high-plains droughts—the "Dust Bowl" years of the 1930s. Hot, dry weather and strong, unending winds wiped out the wheat. Before the farmers came, the topsoil of the plains had been protected by a tough mat of native grasses that turned dry and brown when drought struck but still covered the soil. Plowed land has no such protection, and the topsoil blew away by the thousands of tons. At the height of the drought, in 1934, the blowing soil traveled to the East Coast.

On May 12, 1934, the *New York Times* reported that "a cloud of dust thousands of feet high, which came from drought-ridden states as far west as Montana, 1500 miles away, filtered the rays of the sun for five hours yesterday. New York was obscured in a half-light similar to the light cast by the sun in a partial eclipse."

Thousands of Great Plains farmers were ruined, and millions of acres of farmland were terribly eroded. By the time the rains returned to the plains in the late 1930s, the land of the Dust Bowl states was in fact a great American desert. Under new federal programs, soil conservation practices were undertaken to try to repair the damage that the drought had inflicted on the land. Much of the farmland was returned to pasture, windbreaks of trees were planted, and new farming techniques were developed. But in spite of these changes, another drought that came in the 1950s seriously damaged many parts of the high plains, especially areas that had been converted back from grassland to wheat fields during World War II. And by the mid-1970s, another high-plains drought had struck. Dry weather and high winds in the winter of 1975–1976 caused the worst soil erosion in nearly two decades—4.5 million acres of land in the Great Plains were damaged by wind erosion (Figure 1).

A careful examination of the climatic history of the high plains yields fairly conclusive evidence that the region has had at least eight major droughts spaced about 20 to 22 years apart. There is no way to predict with any certainty that this pattern will continue, but the high plains region has been subject to chronic and severe short-term climatic fluctuations that have had a serious impact on the lives of people who lived in the region. The prudent individual, whether a wheat farmer, an urban planner, or just a resident of the high-plains region, would do well to assume that future droughts will occur and be prepared for them.

*   *   *

Although we have found these examples close to home, they have more than parochial significance: They illustrate some important conclusions about climatic change that are supported by varied and extensive evidence from many other parts of the world. For example:

> During long periods in the past, the worldwide climate has been much warmer than it is now, and tropical or subtropical conditions prevailed in many regions that now have temperate or even frigid climates.

> During other periods, the climate has been much colder, and huge ice sheets crept down from the poles and the mountains to cover much of the present temperate zones.

FIGURE 1 An abandoned farmhouse and windmill standing in the middle of a field of ripe wheat under threatening skies sum up the story of agriculture in the high plains of northeastern Colorado. When the weather is good, it brings rich harvests of wheat; when it is bad, the farmer can be wiped out by hail or drought. (Henry Lansford.)

Climatic fluctuations on much shorter time scales than those of ice ages or global warmings, involving regional variations in rainfall, temperature, winds, and other atmospheric conditions over periods of years or decades, can have substantial impacts on the quality and character of human life.

In its 1975 report *Understanding Climatic Change*, the Panel on Climatic Variation of the National Academy of Sciences pointed out that detailed studies of past climatic changes are a necessary part of any serious attempt to understand and predict the future course of our climate. The panel concluded that "a satisfactory perspective of the history of climate can be achieved only by the analysis of observations spanning the entire time range of climatic variation, say from 10^{-1} to 10^9 years"—that is, from one year to one billion years.

What do we know about past climates, and how have we learned it? Obviously, the further back we go, the less certain our knowledge becomes. For about the past 30 years, we have reasonably good instrumental data from surface weather stations and upper-air balloon soundings in many parts of the world; we also have satellite observations for the last decade or so. For the past 100 years, many nations have operated national weather services, and systematic weather records of varying reliability exist for many locations. For the past 1000 years, historical records of various sorts can be used to piece together a climatic chronology, though much of it is developed by inference and intelligent guesswork. Although some historical sources go back as far as 3000 BC, the majority of the evidence for climatic conditions more than 1000 years before the present is geological and paleontological and often is very approximate.

Let's begin by considering what we know about climates of the past, without reference to the nature of the evidence. Then we will look at some of the methods that have been used to reconstruct the earth's climatic history.

To begin with, for the greater part of its several billion years of existence (sometimes called geological time, by contrast with the brief span of human existence on earth), our planet seems to have had a comparatively warm climate. During what British climatologist C. E. P. Brooks called "the normal climate of geological time," the polar regions appear to have had a temperate climate, and today's temperate zones had the sort of climate that we characterize today as subtropical or tropical. Palm trees were common over the present United States; giant ferns grew in the areas that are now the Dakotas.

But this genial "normal" climate has been interrupted periodically by glacial periods such as the one from which we emerged about 10,000 years ago. Some climatologists like to say that we are still in an ice age on the geological time scale, even though we are going through an interglacial period, when the continental ice sheets have retreated poleward. More of the earth's surface is covered with ice now than was ice-covered during many hundreds of millions of years. Today, both the North and South Poles have extensive ice caps, but during most of the earth's history the poles have been free of ice.

Nevertheless, our present climate is mild by comparison with conditions that have prevailed during most of human life on earth—the last million years or so. As a panel of government scientists reported recently, "Not only is our present-day climate much warmer than the average of the past several centuries, but the latter, in turn, has been much warmer (and more ice-free) than the average of past million years" (Interdepartmental Committee for Atmospheric Sciences, 1974).

The Last 1,000,000,000 Years

Although the evidence for climatic fluctuations on time scales of hundreds of millions of years lacks continuity and detail, the greater part of the last billion years seems to have been favored by the warm ice-free conditions that Brooks called the normal climate of geological time. There is reasonably good geological evidence that this warm climate was interrupted by two glacial ages before the Pleistocene Epoch of the last million years. About 300 million years ago, Southern Hemisphere ice sheets covered parts of present-day South America, Africa, India, Australia, and Antarctica. These land masses apparently were joined together then in a "supercontinent" that the geologists call Pangaea. Evidence of another period or periods of extensive glaciation between about 800 million and 600 million years ago has been found in Greenland, Africa, Australia, and Asia. Although other glacial periods may have occurred before these, the evidence is very dim for such distant times.

The Last 100,000,000 Years

From 100 million to 65 million years ago, the earth was much warmer than it is today, and there were no polar ice caps. About 55 million years ago, polar ice began to grow, and by 25 million years ago the temperature was much cooler. About 10 million years ago, the temperature dropped even more, and both Antarctic ice and Northern Hemisphere mountain glaciers grew substantially. The cooling trend continued into the long cold spell that reached its most frigid extremes during the glaciations of the last million years. This cold period has not yet ended. Our present interglacial conditions are considerably warmer than most of the last million years, but we are still in an ice age by comparison with the climate that existed between 50 million and 100 million years ago.

The Last 1,000,000 Years

For at least the past 1 million years—the Pleistocene Epoch—the earth's climate has been cold most of the time, swinging through a series of glacial and interglacial episodes during which Northern Hemisphere ice sheets alternately advanced and retreated as temperatures dropped and rose in both hemispheres (Figure 2). For many years, geologists had identified only four major glacial episodes during this long Pleistocene cold spell. But more recent research indicates that there were probably at least seven transitions from full glacial to full interglacial conditions, recurring about every 100,000 years.

The Last 150,000 Years

The last warm, ice-free interglacial period comparable to the last 10,000 years began some 125,000 years ago and lasted about 10,000 years. Starting about 115,000 years ago, the temperature began to decline, and by 75,000 years ago the climate was intensely glacial again. Like our present interglacial period, this warm period of 125,000 years ago, known as the Eemian Interglacial, began with an abrupt (at least in geological and climatological terms) change from glacial to interglacial conditions.

The Last 25,000 Years

The major climatic event of the last 25,000 years was the advance and retreat of the great continental ice sheets that marked the most recent glaciation of the Pleistocene Epoch—the name that geologists have given to the million years or so that preceded the last 10,000 years, which is known as the Holocene Epoch. The Pleistocene was a time of cold, and the winters brought more snow than the summers could melt away. On high mountains and in the northernmost regions of North America,

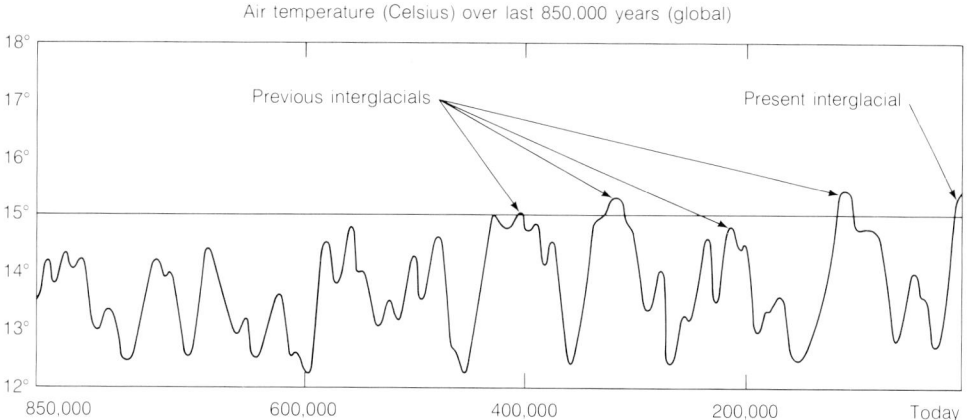

Air temperature (Celsius) over last 850,000 years (global)

Previous interglacials

Present interglacial

Years before present

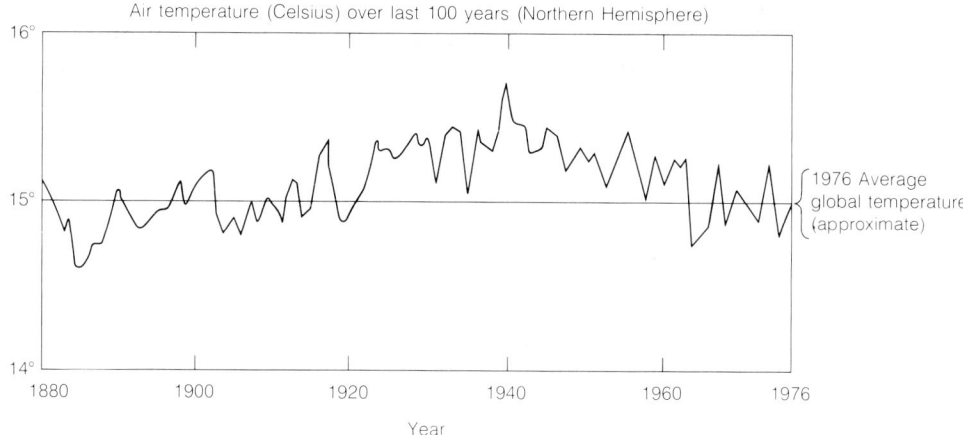

Air temperature (Celsius) over last 100 years (Northern Hemisphere)

1976 Average global temperature (approximate)

Year

FIGURE 2   These two graphs of global temperature fluctuations show that the earth's climate is variable on both long and comparatively short time scales. (National Center for Atmospheric Research, from National Academy of Sciences and other sources.)

Europe, and Asia, the snow grew into great ice sheets that spread under their own weight. In the Southern Hemisphere, there was little glaciation, as there was little land on which the ice sheets could build up and spread out. But in the Northern Hemisphere, the ice sheets grew, reaching out in the coldest periods and retreating temporarily when the cold grew less severe.

The ice spread southward from Scandinavia, reaching west to join a glacier that covered the British Isles and east to merge with the Siberian ice. The mountains of Europe spawned glaciers that reached down the valleys into the lowlands. In North America, ice covered present-day Canada and reached southward across what is now the United States to the Missouri and Ohio River valleys.

The last period of extensive continental glaciation during the Pleistocene occurred between 22,000 and 14,000 years ago, then the ice sheets began to retreat. By 8500 years ago, the ice had withdrawn in Europe to about its present extent, and the North American ice had shrunk to about its present state by 7000 years ago.

The period from 7000 to 5000 years ago was warmer than our present time, but the last 5000 years or so has been marked by generally declining temperatures, with exceptionally cold intervals about 2800 and 350 years ago.

The Last 1000 Years

About 1000 years ago, according to Lamb, the earth was in a dry, warm period, during which the Atlantic Ocean and the North Sea were almost free of storms. This was the time of the great Viking voyages, when Norse adventurers colonized Iceland and Greenland and visited North America. Vineyards flourished in England, indicating high summer temperatures and mild Mays and Septembers. However, there were some frosts in the Mediterranean region and occasional freezing of rivers such as the Tiber at Rome and the Nile at Cairo, suggesting that a shift in the pattern of large-scale weather systems was bringing cold Siberian air further south than before.

By AD 1200, the benign climate in western Europe had begun to deteriorate, and climatic extremes characterized the next two centuries. Great floods and droughts occurred, along with remarkably severe winters as well as remarkably warm ones. The Viking colonies in Greenland and Iceland began to decline as ocean ice increased and the weather grew colder and more stormy. Both English vineyards and vineyards on the Continent fell on hard times.

After a partial return to more favorable conditions from around 1400 to 1550, the climate grew colder again, and the 300-year cold spell known to climatologists as the "Little Ice Age" began. This period, from around 1550 to 1850, was marked by severely cold weather. European glaciers grew, and the River Thames at London began to freeze over with greater frequency than before. Farming became impossible in many parts of the Alps as well as in northern regions such as Iceland and Norway. Vineyards disappeared from England.

The year 1816 was one of the worst of many bad years of the Little Ice Age. In much of Europe, the winter was wet, the spring was cold and late, and the summer was rainy. The wheat harvest was very bad, and famine spread across England, France, and Germany. In the United States, 1816 came to be known as "the year without a summer" or "eighteen hundred and froze-to-death." In the northeastern United States, there was heavy snow in June and frost in July and August. The corn crop was virtually wiped out in northern New England, and there was widespread hunger and even starvation during the bitter winter that followed.

But after the middle of the nineteenth century, the cold temperatures of the Little Ice Age began to moderate, and soon the long warming trend of the first part of the twentieth century had begun.

The Last 100 Years

The aspect of the last century's climate that has provoked the most discussion, in both scientific circles and the news media, is fluctuations in temperature (see Figure 2). Most climatologists agree on one reasonably well-documented fact—much of the Northern Hemisphere cooled off between the 1940s and the late 1960s. The decrease in the average temperature was slight—less than half a degree Celsius (less than 1 degree Fahrenheit) for the Northern Hemisphere as a whole and be-tween 2 and 3 degrees for some northern locations such as Iceland. This cooling, which became more pronounced after 1960, followed a long global warming trend that began in the 1880s.

Although an average temperature change of less than a degree may seem insignificantly small, Lamb has pointed out that the length of the growing season in England increased by two or three weeks during the warming period and dropped back by about two weeks between 1945 and 1970. In addition to such direct effects of the cooling, some scientists see evidence that it caused greater climatic variability—more floods, droughts, and freezes like those that plagued farmers in 1972 and 1974. By 1970, it was not clear whether or not the cooling trend was still in progress, and by the mid-1970s some climatologists saw indications that, at least in the North Atlantic region, the temperature was starting to rise again.

*   *   *

EVEN WHEN IT IS STATED in terms as broad as the ones we have used, this climatic chronology is not built on unequivocal knowledge of our planet's past. We have summarized some features of past climates on which many climatologists, geologists, and other scientists have reached at least a rough consensus, but much uncertainty and controversy remain over climatic variations of the past.

How have we gained even the partial knowledge that we now possess about climates of past centuries and millennia? Systematic measurements of winds, temperature, humidity, and pressure at various levels of the atmosphere have been made for only the last 30 years or so, and organized large-scale observations from surface weather stations go back only a century, so how can we say with any assurance at all that the climate of the last billion years was generally warm, with comparatively brief interruptions by glacial epochs that spanned the last million years and earlier cold periods that occurred about 300 million and 600 million years ago?

Most of our knowledge of climatic conditions more than a century ago comes from two kinds of sources. Historical techniques of climate reconstruction draw on a variety of old records and chronicles that do not necessarily describe weather and climate per se but often deal with crop yields, harbor conditions, and other matters that are affected by climatic conditions or weather extremes. Paleoclimatology, which includes both biological and geological methods of climate reconstruction, draws on proxy records built up by various natural recording systems, such as growth rings in tree trunks, layered lake-bottom sediments, geological features created by the advance and retreat of glaciers, and core samples taken from polar ice sheets that have built up over many centuries.

Although some isolated records of noninstrumental weather observations date back as far as the fourteenth century, the idea that systematic observations of the atmosphere could lead to an understanding of the principles that govern its behavior really emerged in the springtime of modern science—the last half of the seventeenth century. The "new scientists," who followed science as a vocation and pursued it with zeal, rather than dabbling in it like the gentlemen-scholars of earlier years, had appeared and were pursuing knowledge along the lines laid down by Francis Bacon. In his *Novum Organum*, published in 1620, Bacon had proposed a new empirical approach to scientific knowledge, rejecting the reliance on classical sources and deductive reasoning that had been revered by earlier scholars. Bacon believed that the key to understanding the physical world lay in observing natural phenomena and assembling masses of empirical data, collected by many observers, from which generalizations and fundamental principles could be induced.

A good example of this approach to the study of weather and climate is seen in twelve sheets of notes, entitled "Scheme of the Wind and Weather at Llanberis," that are preserved in the National Library of Wales. These notes record daily weather observations made between March 1, 1697, and February 28, 1698, by Thomas Evans, Vicar of Llanberis, Wales, a village at the foot of Mount Snowdon. Evans' notes are succinct but descriptive. March 16, 1697, for example, had a "Cloudy morning and hailstones at noon." The note for January 16, 1698, records "Cold frosty weather with snow," and for January 29, "Rain in the morning and snow at night" (Emery and Smith, 1976).

The Vicar of Llanberis was one of a number of correspondents, most of them clergymen, enlisted by Edward Lhuyd, Fellow of the Royal Society and Keeper of the Ashmolean Museum at Oxford University. A good Baconian in his scientific method, Lhuyd used his network of observers, along with information collected on his own extensive travels, to compile a collection of first-hand empirical data on weather, plants, animals, and geological features, for a study of *The Natural History of Wales*.

As new scientific instruments became available, qualitative weather observations such as those made by Thomas Evans gave way to quantitative instrumental measurements. Since the time of the ancient Greeks, who gave meteorology its name, people had been able to measure only two aspects of the weather—wind direction and precipitation. In the early 1600s the thermometer was invented by Galileo Galilei and the barometer by his pupil Evangelista Torricelli. Now it was possible to measure two of the most important elements of weather and climate: air temperature and atmospheric pressure.

According to Lamb, the earliest instrumental weather records were kept by the Academia Del Cimento at Florence beginning in the 1650s; by an observer named Derham at Upminster, just east of London, in 1697; and by the Kirch family at Berlin between about 1698 and 1770. The French government established a network of weather observations, recorded at police stations, in 1775, and the government of Prussia established a similar system in 1817. The first international network of weather stations, extending from Greenland to Moscow, was organized between 1781 and 1792 by the meteorological society of the Palatinate of the Rhine, with the active support of the ruling prince. In the American colonies, many prominent men took an interest in the weather. Benjamin Franklin's contributions to meteorology are well known; and George Washington, Thomas Jefferson, James Madison, and John Quincy Adams all kept weather records. On the morning of July 4, 1776, a day when he might have been preoccupied with other matters to the exclusion of meteorology, Jefferson took time to note that the 6:00 AM temperature in Philadelphia was 68 degrees Fahrenheit.

Although they are considerably more valuable than the subjective impressions of observers like Thomas Evans, these early instrumental weather records often are hard to evaluate for trustworthiness and comparability with modern data. Efforts to reconstruct the climate of more than a century ago have also been based on other historical sources, such as allusions to weather conditions and events in documents that are devoted primarily to different matters.

Lamb has found abundant manuscript information about the character of particular months and seasons, especially those of dramatic character, from very early historical times. Using state, local, monastic, manorial, and family chronicles, as well as personal diaries and other historical sources, Lamb and his colleagues have reconstructed a remarkable chronology of British and European climate over the past 1000 years. However, as Lamb himself points out, meteorological allusions in nonmeteorological documents tend to emphasize anomalies rather than unremarkable periods of near-normal weather. The climatic historian must be very careful not to give undue weight to local or short-term weather extremes and must take care to avoid overemphasizing them as evidence of climatic trends.

Another kind of historical source that is useful in climatic reconstruction is the continuing record of recurring events that are affected by weather conditions but that are noted routinely whether those conditions are remarkable or unremarkable. For many European locations, the wine harvest is a prime source for the climatic historian.

The study of climatic conditions on the basis of the dates of the ripening of plants and crops is known as phenology. As explained by one of its leading contemporary practitioners, the French historian Emmanuel Le Roy Ladurie, the principle of the phenological method is very simple:

> The date at which fruit ripens is mainly a function of the temperature to which the plant is exposed between the formation of the buds and the completion of fruiting. The warmer and sunnier this period is, the swifter and earlier the fruit (or crop, in the case of cultivated plants) reaches maturity. Conversely, if these months have been cold, cloudy, and dull, ripeness and harvest are held back. There is a close correlation, which has been verified with great accuracy in many plants, between the total temperatures of the vegetative period and the dates of blossoming and fruiting. These dates are thus valuable climatic indicators.

Le Roy Ladurie goes on to point out that "for the historian the field of research here is at once limited: the *ancien regime* left little evidence about the dates of the lilac and the rose." However, there is one plant whose ripening has been noted every year for many centuries all across Europe—the grape. In hundreds of cities and towns in Champagne and

Burgundy, along the Rhine and the Mosel, and in many other wine-growing districts of Europe, the date of the wine harvest traditionally has been fixed by proclamation, on the basis of a decision made by a panel of local experts, and documented in public records.

This record, which extends back as far as the fourteenth century in some regions, has a great advantage: It continues into modern times, and thus it can be calibrated against instrumental records of temperature. Another French historian of climate, Garnier, has compared nineteenth-century wine harvest dates in Argenteuil, Dijon, and Volnay with average April–September temperatures recorded during the same period by the Paris Observatory and has established that a consistent relationship exists between these two variables.

The first scientist to bring together large amounts of data on wine harvests as sources of climatic history was A. Angot of the French Central Meteorological Office. In 1883, he published an article based on French, German, and Swiss wine-harvest records that went back as far as the fourteenth century. However, as Le Roy Ladurie puts it, the phenological data are "copious for the nineteenth century, abundant for the eighteenth, adequate for the seventeenth, and fairly rare before that." Nevertheless, Angot and his successors have used the European wine harvests to fill many gaps in the climatic record over the last several centuries.

Phenological data from other plants in various parts of the world have also been used in efforts to reconstruct climate. For example, in Japan the date that the blossoms open on the Emperor's cherry trees has been recorded for hundreds of years. In recent years, the tendency has been toward later blooming dates, suggesting a cooling trend. Stephen Schneider of the U.S. National Center for Atmospheric Research has pointed out, however, that the trees might bloom later not because of cooler weather but because they have been damaged by Tokyo's notorious air pollution. He also points out that the earlier records, which are for the cherry trees at the ancient capital of Kyoto, indicated the date that the Emperor made his annual excursion to view the cherry blossoms rather than the date that they actually bloomed.

Le Roy Ladurie has also cautioned against overly simple interpretations of phenological indicators. In scanning long-term wine-harvest records from all over Europe, he has determined that they are generally consistent in their indications of warm or cool summers, but sometimes the harvest was considerably later at certain vineyards, even compared to others nearby. This anomaly is usually indicative of a change in wine-growing practice rather than in climate. A later harvest produces grapes that are richer in sugar, resulting in better wines that will keep their quality longer. When the demand for such wines increased in a particular region, the harvests occurred later in the year, and thus the harvest date reflected economic as much as atmospheric fluctuations. The important point to note here is that, with phenological methods, as with most other techniques for reconstructing climate from sources other than instrumental weather records, the data must be examined very carefully to identify any factors other than climatic influences that may account for fluctuations.

Another biological technique of climate reconstruction uses a record that is actually preserved in the structure of the organism itself. In the early years of the twentieth century, Andrew E. Douglass, a solar astronomer at the University of Arizona, suggested that the annual growth rings of trees growing in dry climates, such as that of the southwestern United States, could yield information on year-to-year climatic fluctuations.

Tree rings form because the layer of plant tissue just under the bark produces large, thin-walled wood cells at the beginning of each growing season. This results in a sharp line of demarcation between the last wood formed during one year and the first wood of the next year. If the tree is cut down, these layers appear as rings on the top of the stump. Douglass hypothesized that for trees that are under climatic stress, the relative thickness of the growth rings from year to year would reflect variations in the factor that induced the stress. For a ponderosa pine growing on a dry mountain slope in southern Arizona, for example, the growth ring would be narrow in a drought year, wider in a year of adequate rainfall.

Douglass spent many years collecting cross sections and radial core samples, taken by inserting a hollow drill into the tree trunk, from ponderosa pines and Douglas firs in the southwestern United States. Although he lacked long-term weather records for the region, he was able to correlate the relative thickness of the tree rings with periods when rainfall was known to have been adequate or low. By working with timbers taken from prehistoric Indian dwellings, matching their later growth rings with earlier sequences in living trees, he was able to extend his records hundreds of years into the past. This technique was used to establish the times that the ancient Indian settlements were built and occupied.

The science of using tree rings to date archaeological evidence came to be known as dendrochronology, and the subdivision dealing with climatic variations as dendroclimatology. The work of Harold Fritts and others who succeeded Douglass has been carried on in the Laboratory of Tree-Ring Research of the University of Arizona. Equipped with high-speed computers and modern techniques of statistical analysis, scientists at this laboratory have correlated the characteristics of tree rings with temperature and precipitation and have produced maps of reconstructed climatic conditions by decades for the southwestern United States many hundreds of years ago. Tree-ring studies of climatic history, using fossil trees as well as living ones, have provided data on climatic fluctuations in Canada, England, Germany, the Soviet Union, and other parts of the world.

Other biological indicators have been used to reconstruct past climates in particular regions. Between 3000 and 1100 BC, for example, bamboo, which needs a fairly warm climate, grew in regions of China some three degrees of latitude farther north than it is able to survive now. Plum trees grew all over China during a warm period before 500 BC but disappeared from north China during a cooling period after AD 900.

However, as Le Roy Ladurie has pointed out emphatically, such evidence should be examined very rigorously before it is accepted as proof of climatic change. He cites the example of a fourteenth-century decline in the output of vineyards north of Paris that had been interpreted by some climatic historians as evidence of a cooling trend. Further research revealed that, as a result of the vast numbers of people killed by war and plague in the mid-1300s, labor had become scarce and expensive. The high cost of hiring laborers to cultivate the vines, rather than a lack of warm summer weather, seems to have been the key factor in the decline in the Paris vineyards. The same sort of simplistic approach has been the basis of some splendid but flawed forays into the study of climatic influences on human history. As Le Roy Ladurie points out, Ellsworth Huntington, a student of climate and history who viewed climate as the primary determinant of human character and behavior, attributed the great Mongol migrations of the thirteenth century to fluctuations in rainfall and barometric pressure in central Asia. Then C. E. P. Brooks "carried on the good work" by basing a graph of rainfall in central Asia on the migrations of the Mongols. "What better example," Le Roy Ladurie asks sarcastically, "of a serpent biting its own tail?"

Following his own advice, Le Roy Ladurie carefully cross-checked historical texts and pictorial evidence against each other and against physical evidence in his studies of another important proxy source of climatic data from the past: the glaciers of the European Alps. Valley glaciers in mountain ranges are remarkable integrators and amplifiers of small, long-term trends in temperature. If, over a period of years, slightly more ice melts from the foot of a glacier than is replaced by snow that falls on its upper reaches the following winter, the glacier will start a slow retreat up its valley. If, on the other hand, more snow falls in the winter than melts in the summer over a period of years, the glacier will slowly advance into the lowlands.

The most striking product of Le Roy Ladurie's studies of Alpine glaciers is a set of pictures. They are in pairs, one showing a particular glacier as it was depicted by a painter or engraver a hundred years or more ago and the second a photograph taken from the same viewpoint in the 1960s (Figure 3). One of these engravings shows the Argentière Glacier, in the French Alps near Chamonix, as it appeared in the 1850s. The huge mass of ice looms beyond a fringe of trees and the church steeple and rooftops of the village of Argentière, overflowing the steep valley that rises southward into the mountains. In a photograph taken in 1966 from the same angle, the ice has retreated far up the valley, and the trees have advanced up the lower slopes that were icebound a hundred years before. Similar pairs of pictures show the retreat of other French, Swiss, and Austrian Alpine glaciers during the warming period of the late nineteenth and early twentieth centuries.

Le Roy Ladurie has correlated this pictorial evidence with two other kinds of records: maps and written accounts by glaciologists, who were studying the glaciers themselves, and a variety of other historical records that contain references to the extent and condition of particular glaciers at particular times. For example, on August 4, 1546, an early traveler and student of natural history named Sebastian Münster rode on horseback to the foot of the great Rhone Glacier, below the Furka Pass in the Swiss Alps. Writing in Latin, he described the glacier vividly; here is the English translation:

> On August 4, 1546, as I was riding toward Furka, I came to an immense mass of ice. As far as I could judge it was about two or three pike lengths thick, and as wide as the range of a strong bow. Its length stretched indefinitely upward, so that you could not see its end. To anyone looking at it it was a terrifying spectacle, its horror enhanced by one or two blocks the size of a house which had detached themselves from the main mass. White water flowed out of it, so full of particles of ice that a horse could not ford it without danger.

Translating Münster's pike lengths and bowshots into modern measurements, Le Roy Ladurie, who retraced Münster's journey twice himself, concludes that this description indicates that the Rhone Glacier

FIGURE 3   Top: The Argentière Glacier in the 1850s. Bottom: The Argentière Glacier in 1966. (From *Times of Feast, Times of Famine* by Emmanuel Le Roy Ladurie. Translated by Barbara Bray. Copyright © 1971 by Doubleday & Company, Inc.)

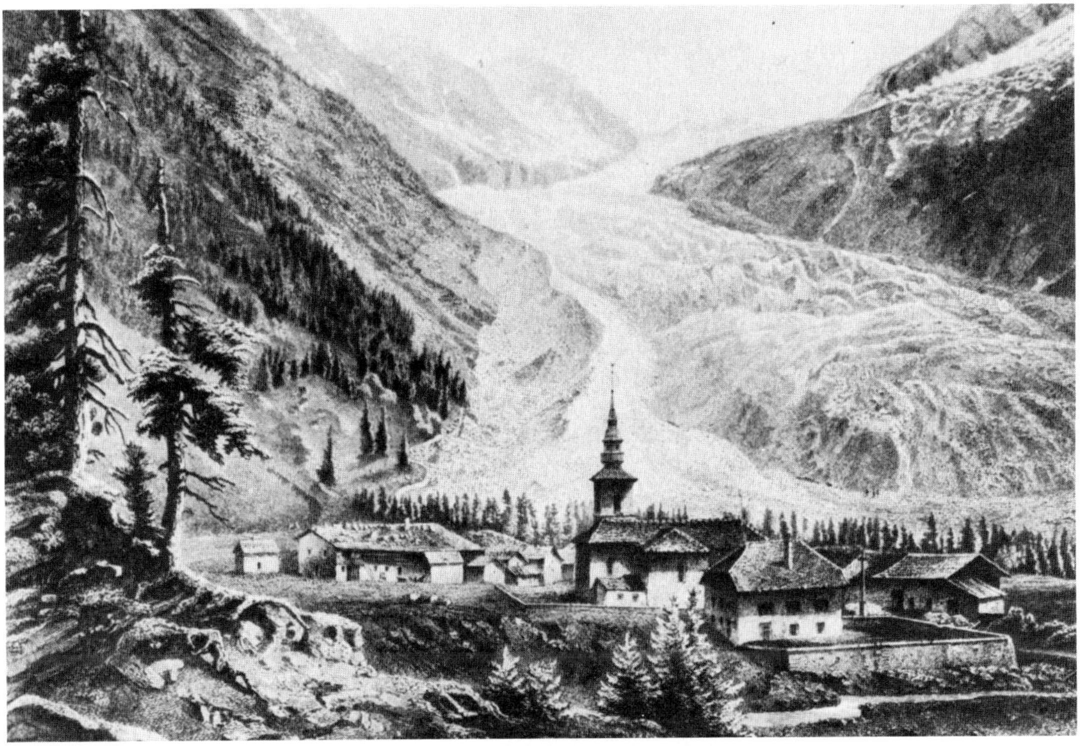

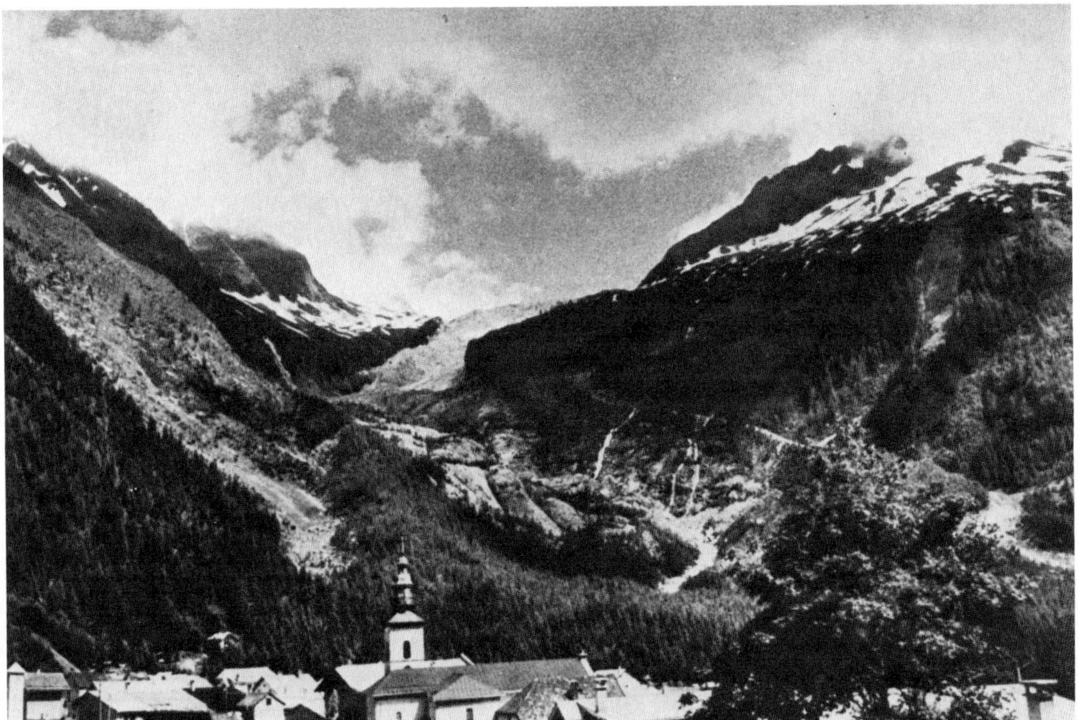

was at least as massive and extensive in the mid-1500s as it was in the late 1800s. Admitting that the evidence is somewhat ambivalent, he also speculates that, when Münster saw the glacier, its extent may have been even greater, as it appears in the earliest known engravings, made around 1700. In any case, the Rhone Glacier was a great deal bigger in 1546 than it is today.

In his studies of glaciers as indicators of long-term fluctuations in temperature, Le Roy Ladurie has been able to draw on extensive glaciological studies that date back to the early nineteenth century. The earliest really compelling evidence for extreme changes of climate in past eras was provided by the striking similarity between boulders and rubble found all over Europe and the moraines and debris left by Alpine glaciers during advances and retreats in recent times.

By the beginning of the nineteenth century, the age and early history of the earth had become subjects of great controversy in scientific and religious circles. As late as the eighteenth century, the Biblical version of the earth's past was accepted. Medieval Jewish scholars had set the date of the Creation at 3760 BC. Around 1650 James Ussher, the Anglican Archbishop of Armagh, Ireland, calculated on the basis of Biblical chronological evidence that the earth was created at 9:00 AM on October 26, 4004 BC. In the late eighteenth century and the early years of the nineteenth, in spite of various heresies proposed by irreverent scientists, most people in the Western world believed that the earth was about 6000 years old, that its plants and animals were the same species that God had created in the Garden of Eden, and that the Biblical accounts of the Creation, the Flood, and the other major events of the Old Testament provided the only reliable and acceptable chronology of the earth's past.

An increasing number of scientists disputed this orthodox view. As early as 1686, the British scientist Robert Hooke concluded that fossil shells found in England but resembling those of contemporary tropical species indicated that the British climate had cooled substantially since the time the shells were deposited. The eighteenth-century French naturalist Georges Louis Leclerc de Buffon proposed that the earth was "a dislodged fragment of the sun, which gradually congealed in the chill of space." By performing laboratory experiments with heated spheres, Buffon calculated that the earth was 74,832 years old and that it had been cool enough to support life for exactly 40,062 years.

In 1785, the Scottish naturalist James Hutton proposed that natural geological processes, such as mountain building and erosion, had been working at about the same rate throughout the earth's history—the "uniformitarian principle." To shape the earth's surface into its present state, Hutton concluded, these processes must have been at work for many millions of years rather than a few thousand. Although Hutton's theories were rejected by some scholars, they were taken up and advanced by others. In the 1830s, the British geologist Charles Lyell built on Hutton's work to shape the basis of modern earth science in his three-volume *Principles of Geology*.

This was the setting in which geological evidence for past glacial periods was first interpreted to prove that ancient climates must have been much colder than modern ones. In Switzerland, the evidence was fairly straightforward. Boulders and rubble found at lower elevations some distance north of the Alps were so similar to those found near existing mountain glaciers that it seemed clear that the Alpine glaciers had once reached out far beyond their present limits, carrying great masses of rock with them. The gouged and polished surfaces of rock that lay between the glaciers and distant deposits of boulders and rubble also supported this theory—the rock had obviously been marked by the force of great masses moving across it.

But a prehistoric advance of Alpine glaciers could not account for similar debris found far from any existing glaciers—in Norway, England, and northern Germany, for example. If this debris had been heaped up by glaciers in regions where none exist in modern times, then radical changes in climate must have occurred at some past time. Noah's flood was proposed by some as the great force that had scattered the rocks about, but the scraped and polished bedrock did not fit this explanation.

In 1821, a Swiss engineer, Ignaz Venetz, read a paper before the Helvetic Society in Lucerne, suggesting that much of Europe must have been covered at some past time by huge glaciers reaching northward from the Alps. In 1824, Jens Esmark proposed that the fjords of his native Norway had been carved by glaciers that pushed down from the Norwegian mountains. In 1832, a German professor named Bernhardi published a paper in which he presented for the first time the theory that Europe had been covered with a huge ice sheet extending from the Arctic regions to the Alps.

Professor Bernhardi was an obscure scholar who taught at a school of forestry in a small town in Thuringia, and his theory attracted little attention from the rest of the world. But when a vigorous and outspoken young Swiss scientist, Louis Agassiz, developed a similar theory of European glaciation, it got a great deal of attention. Skeptical at first of the ideas of the early Swiss glaciologists, Agassiz performed some experiments on the movement of glaciers; went on a field trip through the Rhone Valley with Jean de Charpentier, a colleague of Venetz; and examined the work of several other scientists who had been studying glaciers. He became convinced that glaciers had done the work of transporting boulders and rubble and scouring the bedrock, and in 1837 he presented the concept of an ice age of European glaciation in a talk to the Helvetic Society. In a book published in 1840, he maintained that great sheets of ice, which resembled those that existed in Greenland, had once covered all of the countries where unstratified gravel, or glacial drift, is found. After emigrating to the United States in 1846 and joining the faculty of Harvard University, Agassiz continued his glacial studies in North America. Although Agassiz won wide acceptance for the ice-age theory, it was still a source of both religious and geological controversy as late as the end of the nineteenth century.

The observations on which Agassiz and his colleagues built their theory of glaciation were based on evidence found at the earth's surface. But in recent years, scientists who are looking for evidence of climatic change in the geological and glaciological record have had a valuable new tool—the core drill. This device uses a hollow-stemmed drill to probe thousands of feet into ocean beds or ice caps, bringing up a cylindrical sample of the sediments or ice layers that have accumulated over many centuries.

Ocean-bed core samples are the most important source of the paleo-climatic data used in a research project known as CLIMAP, Climate Mapping and Prediction, a joint effort by scientists from four U.S. universities and several European universities. The goal of the CLIMAP scientists is to map the earth's climate at certain critical times in the past to provide a foundation for efforts to predict future climate. Maps are being constructed that show ocean temperature and other climatic factors at carefully selected times in the past—for example, 18,000 years ago, when the most recent glacial period was at its peak (Figure 4). Other times for which climate maps will be constructed are the warm period that occurred about 6000 years ago; a previous warm period that occurred 120,000 years ago, before the onset of the last ice age; and a period about 700,000 years ago, when the earth was about to enter an earlier ice age.

Plankton—tiny temperature-sensitive life forms that drift along in the ocean—are a key element in the CLIMAP reconstructions of past climates. Some species are able to survive only within a narrow range of temperatures, with the range varying from one species to another. Thus the distribution of different kinds of plankton in a sample of ocean water is an indicator of the temperature of the water.

In the ocean, plankton exist in huge numbers, and they do not live very long. Thus a constant rain of tiny bodies falls constantly on the ocean floor, building up in layers over the years and centuries. A core sample from the bottom of the ocean contains cross-sections of these layers; by determining when a particular layer was deposited and analyzing the distribution of plankton species in that layer, we can determine what the ocean temperature was at the distant time that the creatures in that layer drifted down to the sea floor.

Mapping ocean temperatures 18,000 years ago is not that simple, of course, but the technique just described is a basic tool of CLIMAP. One kind of plankton known as foraminifera—foram for short—is particularly useful. A foram about the size of a pinhead, *Neogloboquadrina pachyderma*, has a spiral shell that turns one way when the prevailing water temperature is colder than 7 degrees Celsius (45 degrees Fahrenheit) and the other way when higher temperatures occur. Another species of foram, *Globorotalia menardii*, is found only in water warmer than 10 degrees Celsius (50 degrees Fahrenheit).

Analysis of the relative amounts of two isotopes of oxygen in the shells of forams also provide a clue to past climatic conditions. The two forms of oxygen—oxygen-16 and oxygen-18—are both found in sea water and are absorbed by the foram shells. Evaporation of water from the ocean into the atmosphere removes more oxygen-16, leaving a higher ratio of oxygen-18 (sometimes called heavy oxygen) in the ocean. Thus during a glacial period, when more water evaporates, falls as snow, and is locked up in ice sheets, the ratio of oxygen-16 to oxygen-18 in the oceans, and thus in the foram shells, will be low. Such chemical evidence indicates that the extent of ice on the earth was large.

The ratio of oxygen isotopes in ice cores taken from the Arctic and Antarctic ice caps also provides a basis for reconstructing past temperatures. Danish, Swiss, and U.S. scientists have drilled deep into the Greenland ice cap to obtain core samples that give clues to temperatures more than 100,000 years into the past. A 1.5-mile-long ice core from Byrd Station in Antarctica has provided data that go back 75,000 years.

Other paleoclimatic data sources have been valuable in reconstructing climates of the past. Fossil pollens retrieved from peat bogs in North America and other mid-latitude locations have provided information on temperature, precipitation, and soil moisture. Lake-bottom sediments, ancient soil deposits, and other natural records also have been useful. But our knowledge of climates of the past is still fragmentary and incomplete, and it seems clear to most climatologists that we will have a better chance of understanding and predicting future climatic fluctuations and changes if we continue to improve our knowledge of past ones.

FIGURE 4 This global map of sea-surface temperatures, ice extent, ice elevation, and continental albedo (the relative amounts of solar radiation absorbed and reflected by the earth's surface) represents a geological reconstruction of conditions during August, 18,000 years before the present, during the peak of the last ice age. (CLIMAP Project.)

A Snow and ice (ice elevation indicated); albedo over 40 percent.

B Sandy deserts, patchy snow, and snow-covered dense coniferous forests; albedo 30–39 percent.

C Loess, steppes, and semideserts; albedo 25–29 percent.

D Savannas and dry grasslands; albedo 20–24 percent.

E Forested and thickly vegetated land; albedo below 20 percent.

F Ice-free ocean and lakes with isotherms of sea-surface temperature in degrees Celsius; albedo below 10 percent.

This view of the earth and its cloud cover, photographed from a satellite in an orbit 23,000 miles from the surface of the planet, shows the complexity of the atmospheric patterns that bring us our weather and climate. (National Aeronautics and Space Administration.)

CHAPTER 4

The Weather and Climate Machine

F ROM THE standpoint of a human observer caught in the middle of a Great Plains blizzard, a Big Thompson flood, or a Dust Bowl drought, the behavior of the atmosphere may appear random, chaotic, and even malevolent. But the same regular and dispassionate physical principles that govern simpler natural systems also apply to the massive and complex system comprised by the sun, the atmosphere, the oceans, the land and ice masses, and the life forms on the face of our planet. Although the processes and interactions that operate in this system are very complex, most atmospheric scientists believe that the behavior of the weather and climate machine can eventually be comprehended in a quantitative and reasonably precise way.

Over the last century, meteorology—the science of the weather—has evolved from a discipline based almost entirely on observation, description, and classification of weather phenomena to one in which a quantitative and analytical understanding of the principles that govern those phenomena has emerged gradually. Although our main concern in this book is with climatic change, which is not very well understood, rather than weather, about which we have a good deal of knowledge and understanding, both weather and climate are defined by certain fundamental properties of the atmosphere—primarily temperature, precipitation, and winds. To consider mechanisms of climatic change intelligently, it is necessary to have some knowledge of how the weather works. The reader who already has such knowledge may skip over the next few pages.

Weather happens in the troposphere, the layer of the atmosphere nearest to the earth's surface. The troposphere extends up about 16 kilometers (10 miles) from the surface at the equator and thins down to about 8 kilometers (5 miles) at the North and South Poles. Although the troposphere is the thinnest layer of the atmosphere in vertical extent, it is much more dense than the stratosphere and other layers that lie above it, comprising 80 percent of the total weight of the atmosphere and containing nearly all of its water vapor. The troposphere normally is warmest near the earth's surface and grows cooler with increasing altitude.

The sun supplies the energy that drives the weather and climate machine. This energy is tremendous; it has been estimated that the solar energy that falls on our planet in a single week is greater than the total energy produced by all the coal, gasoline, and other fuels that the human race has ever burned.

When it reaches the earth's atmosphere, some of this short-wave radiant energy, or sunshine, is reflected and scattered back into space. But part of it passes through the atmosphere. When it reaches the earth's surface, another fraction of the energy is reflected upward, but more is absorbed by the land and the oceans, which then release much of it back into the atmosphere as long-wave radiant energy, or heat. Part of the solar energy that falls on the oceans, plants, damp soil, and other surfaces that contain moisture, does the work of evaporating water and enters the atmosphere as latent heat—energy that is released again when the water vapor condenses back into cloud droplets.

In the long run, the earth radiates about as much energy back into space as it receives from the sun. If it didn't, our climate would grow markedly hotter or colder over the years, instead of fluctuating over a range of only a few degrees of temperature. The earth absorbs more energy near the equator, where the sun's rays strike its surface directly, and less near the poles, where the solar radiation strikes at a shallow angle. As all parts of the globe radiate energy back into space at roughly the same rate, energy must somehow be transported from the equatorial to the polar regions. The basic mechanism for this energy transfer is convection, in which a fluid—in this case air—is heated, and the heat is transported by movement of the fluid. Atmospheric convection is the driving force behind the air motions that we call wind.

Wind is more than just a single feature of local weather. It occurs on every level from large-scale prevailing air currents, such as the trade winds and westerlies, down to local winds, such as sea breezes. There are three scales of atmospheric motion, all of which play major roles in determining what kind of weather occurs at a particular time and place.

The largest scale of atmospheric motion, the general circulation, is the basic mechanism by which energy is transported through the atmosphere from the equator toward the poles. If the earth did not rotate, if it had a uniform surface—all ocean or all bare, level land—and if solar energy arrived with equal intensity all around the equator, the general circulation would be very simple. Warm, light air would rise over the tropics. Cool, heavy air would move in from north and south to replace it. A pattern of circulation would be established with warm, high-tropo-

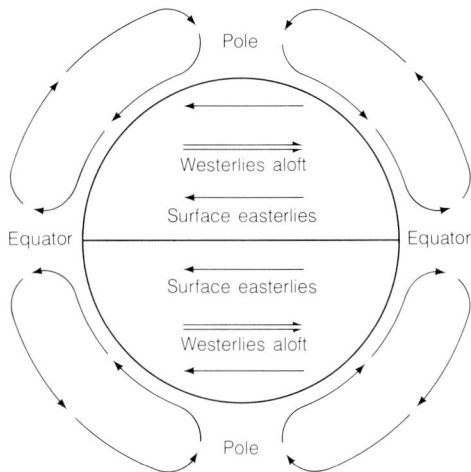

FIGURE 1   Hadley's model of the global atmospheric circulation was basically correct, but vastly oversimplified. (From M. Neiburger, J. G. Edinger, and W. D. Bonner, *Understanding Our Atmospheric Environment*, W. H. Freeman and Company, San Francisco. Copyright © 1973.)

spheric air moving toward the poles and cool, low-level air moving back toward the equator. And this pattern does exist for some distance north and south of the equator. The air moves upward, poleward, downward, and then equatorward, in patterns known as Hadley Cells, for George Hadley, the Englishman who first identified them in 1735 (Figure 1).

But the real atmosphere is not as simple as Hadley believed. The earth does rotate, its surface is divided between oceans and irregular land masses, and the input of solar energy to different regions varies with the days and the seasons. As a result, Hadley Cells give way to more complex patterns of atmospheric circulation north and south of the tropics (Figure 2). The general circulation begins along the equator, where warm air rises in strong updrafts, creating a belt of low atmospheric pressure. As the rising tropical air starts to accumulate at high altitudes, it diverges, spreading toward the north and south. At latitudes of about 25 to 30 degrees in both the Northern and Southern Hemispheres, part of the air starts to descend. This produces the subtropical high-pressure belts called the horse latitudes (because sailing ships carrying horses to the West Indies were sometimes becalmed there and forced to throw the horses overboard as food and drinking water ran short). As the descending air reaches the surface, part of it goes back toward the equator, as Hadley proposed. But part continues toward the poles, creating two great bands of surface winds.

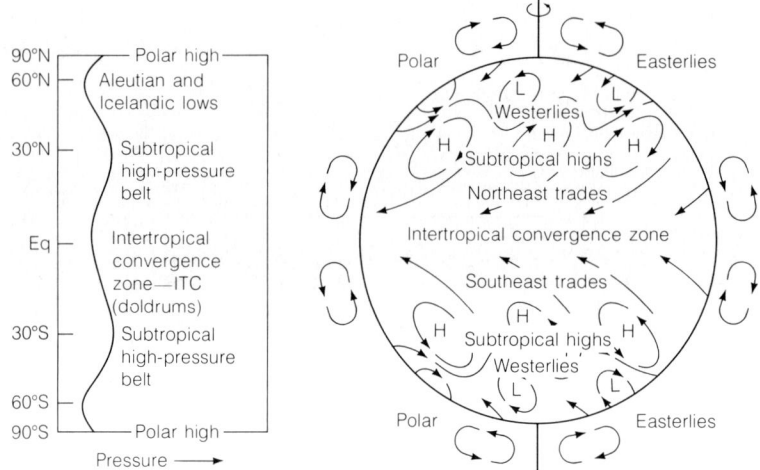

FIGURE 2 This schematic representation of features of global circulation shows many of the complex motions that Hadley neglected. (From M. Neiburger, J. G. Edinger, and W. D. Bonner, *Understanding Our Atmospheric Environment*, W. H. Freeman and Company, San Francisco. Copyright © 1973.)

Through an effect known as the Coriolis force, the earth's rotation twists these winds to the right in the Northern Hemisphere and to the left in the Southern. Thus the winds blowing toward the equator come from the northeast in the Northern Hemisphere and from the southeast in the Southern Hemisphere. In the days when trading ships sailed the tropical oceans, sailors named these steady and dependable air currents the trade winds; meteorologists know the region where they come together along the equator as the intertropical convergence zone, or the doldrums.

The winds that move poleward from the horse latitudes are also twisted by the earth's rotation, and they become the antitrades, or prevailing westerlies. The normal west-east movement of weather systems across the United States and western Europe is produced by the prevailing westerlies. In the Southern Hemisphere, at latitudes around 40 degrees, there is not much land to drag at the westerlies and slow them down, and the strong winds that blow over the oceans there have given this region the mariners' nickname, roaring forties.

When the remaining high-level, poleward-moving air cools off and descends over the north and south polar regions, it produces the polar easterlies, bitterly cold winds that sweep down over North America and northern Europe and Asia in the winter.

The dominant vertical and horizontal motions of the general circulation create weather patterns that prevail in certain regions of the world.

For example, equatorial Africa and South America are known not only for their lack of winds but also for regular and heavy rainfall. The Amazon River basin of South America, straddling the equator, often has close to 450 centimeters (180 inches) of rain a year. This heavy rainfall is caused by one of the fundamental principles of atmospheric behavior: Rising air produces clouds and precipitation. As air near the surface of the earth is heated, it expands and grows less dense. This causes it to rise, and it continues to expand as it rises to higher altitudes, where the atmosphere is less dense. One of the basic laws of physics is that an expanding gas grows cooler, and cool air can hold less water vapor than warm air. As a result, water vapor in the rising air begins to condense into cloud droplets, which eventually form raindrops. Thus a region of predominantly rising air, such as the one along the equator, is very rainy.

Conversely, when the vertical motion of the air is consistently downward, fair weather is the rule, and rain falls infrequently. The deserts of the North American Southwest, the Sahara, and many other arid areas of the world are located in the high-pressure belt of the horse latitudes. Of course, many other factors besides the general circulation affect local weather and climate, and not all the areas in the horse latitudes are deserts.

Another cause of large-scale as well as small-scale patterns of prevailing weather is the contrast between the rates at which land and ocean absorb and radiate energy. Land masses heat and cool rapidly, while oceans gain or lose heat slowly; this is what causes sea breezes. The land heats rapidly, causing the air above it to expand. This produces a low-pressure area, and cooler air blows in from the sea to fill it. At night, the land cools more rapidly than the sea, and the wind turns to blow out to sea from the shore.

The world's largest land masses, such as the Asian subcontinent, produce monsoons, winds that are much like local sea breezes except that they are on a larger scale and reverse with the seasons instead of the day and night. In winter, the air over Asia grows cool and dense, and it spreads toward the Indian and Pacific Oceans, producing the winter monsoon and fair weather. When spring comes, the air over the land becomes hotter and lighter, and the wind sweeps inland. The summer monsoon brings in moist air to produce torrential rains—as much as 710 centimeters (280 inches) per year in some mountain areas of India. Monsoons are also characteristic of other regions, such as West Africa, where the Sahel depends on monsoon rains to water its sparse vegetation.

The second scale of atmospheric motion, smaller than the general circulation but still enormous by human standards, is the synoptic, or cyclonic, scale. Motions on this scale are characterized by rotating weather systems, hundreds of kilometers across, known as cyclones and anticyclones. Cyclonic-scale weather systems cause the pattern of alternating good and bad weather that is characteristic of most of the United States, western Europe, and other temperate-zone areas. Cyclones are produced when an upper-air disturbance triggers a reaction between dissimilar air masses in the region that is influenced by both the prevailing westerlies and the polar easterlies.

For example, a great mass of moist, light tropical air may be carried north from the Gulf of Mexico, colliding somewhere over Kansas with a mass of cold, dry, heavy air from Canada. Usually one of these air masses will be moving more rapidly than the other. If the cold air pushes in under the warm mass, lifting it up, the line where they meet is called a cold front. If the warm air mass moves into an area occupied by cold air, the line of collision is called a warm front. In either case, the usual result is stormy weather. Cold fronts often bring thunderstorms, tornadoes, and other violent weather phenomena; warm fronts usually produce cloud cover and steady rain.

Along strong, active fronts, the air currents in the two masses usually move in opposite directions. Twisted by the earth's rotation, these winds may start to rotate counterclockwise, spiraling inward toward a center of low pressure. The counterclockwise weather system, circling around a low center, is a cyclone, sometimes called a low or depression. The clockwise system is known as an anticyclone. These great disturbances, marching along one after another, bring us our normal pattern of several days of bad weather alternating with several days of good weather.

In recent years, meteorologists have found that the cyclonic-scale systems that move through the mid-latitudes are generated and steered to a great extent by strong westerly winds high in the troposphere. During World War II, bomber pilots discovered in the upper-level westerlies high-speed cores of air that have come to be known as jet streams. The polar jet stream, which divides cold polar air from warm tropical air, oscillates from north to south as it circles the globe, forming between one and six long waves, or zigzags. Where the jet stream swings poleward, warm tropical air follows it north; where it dips back toward the equator, it brings cold Arctic air down with it.

In the summer, when the contrast between temperature at the equator and that at the poles is less, the band of upper-level westerlies, sometimes called the circumpolar vortex, moves poleward and grows weaker. In the winter, when the polar regions are colder compared with the tropics, the upper-level westerlies normally move equatorward and grow more vigorous. When the jet stream forms a pattern of deep ridges and troughs and holds it for weeks or even months, the same weather conditions persist, and long heat waves, droughts, or periods of rain or snow occur. This is what meteorologists call a blocking situation, because the pattern of the upper-level winds has blocked the normal west-east progression of cyclones and anticyclones that brings changing weather.

Weather on the third, or local, scale is much less predictable than it is on larger scales. The thunderstorm that rains out a picnic or ball game or brings a flash flood can grow very suddenly. Thunderstorms are produced by rising air and often are associated with cold fronts, although a front is not a prerequisite for thunderstorm development (Figure 3). Upward motion of air, the first requirement for thunderstorms, can be created by local differences in the earth's surface. On a bright summer day, the air over an open field will be heated more than the air surrounding forests. The warm air, expanding and growing lighter, will start to rise. If the air is moist enough, its water vapor will begin to condense into cloud droplets, and the change of state from vapor to liquid will release

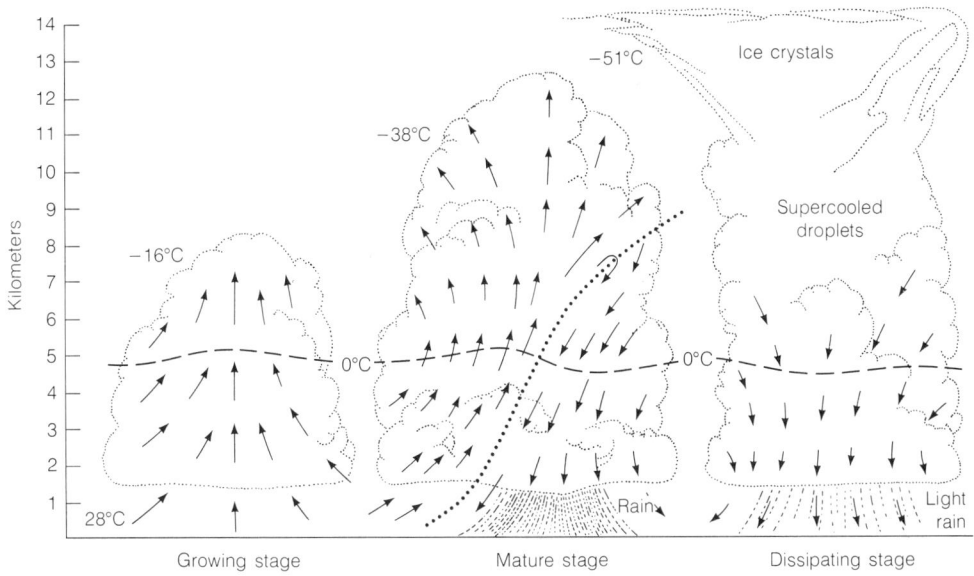

FIGURE 3 Even a thunderstorm, one of the most common weather phenomena, has dynamic and physical features that can be much more complex than this simple diagram of its life cycle indicates. (From M. Neiburger, J. G. Edinger, and W. D. Bonner, *Understanding Our Atmospheric Environment*, W. H. Freeman and Company, San Francisco. Copyright © 1973.)

more heat, strengthening the upward motion. If conditions are right, a cumulus cloud will form and grow into a cumulonimbus, or thunderhead. If the growth process is strong and sustained, the thunderhead can spawn torrential rain, lightning, thunder, hail, and sometimes tornadoes, and can do a great deal of damage during its life cycle of a few minutes or hours.

Weather is a much more complicated and fascinating matter than this brief review may suggest, and the reader who is inclined to pursue the subject can find many sources of solid and interesting information about weather in any good library. Our main concern here is with atmospheric behavior on a longer time scale, and, with apologies for treating a subtle and complex subject in such a cursory way, we must return to questions of climate.

There is no single theory, or even a combination of a small number of theories, that satisfactorily explains climatic variation on all time scales. But a number of factors clearly are important, either in terms of their correlation with past climatic variations or because tentative theories about climate mechanisms indicate that they probably play significant roles. In the past, some climate theoreticians attached themselves to one or another of these factors and worked very hard to prove that it was the primary determinant of climatic change. The British climatologist C. E. P. Brooks, who was a highly influential figure in the development of climate theory in the period between the two world wars, believed strongly for many years that geographical changes—mountain building and erosion and changes in the distribution of land and oceans—were the dominant influences on climatic change. But in the preface to his book *Climate Through the Ages*, first published in 1922, Brooks quoted Kipling's lines:

> There are nine and sixty ways
> Of constructing tribal lays,
> And every single one of them is right.

"There are at least nine and sixty ways of constructing a theory of climatic change," Brooks observed in 1949, "and there is probably some truth in quite a number of them. The greatest extremes of climate are not to be attributed to the abnormal development of one factor, but to the co-operation of a number of different factors acting in the same direction."

Keeping this wise admonition in mind, and also keeping in mind that climatic changes and fluctuations on different time scales may be influenced by very different factors, let us look at some current views of the forces that steer the climate machine. We will not consider nine and sixty

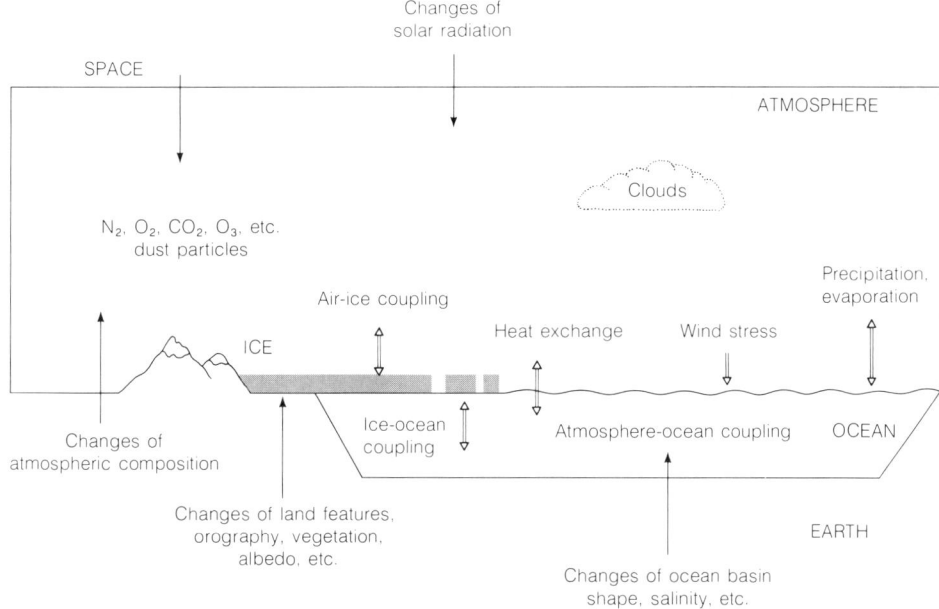

FIGURE 4   A schematic illustration of the main components of the coupled atmosphere-ocean-ice-earth climate system. Full arrows (⟶) indicate external influences; open arrows (⟹) indicate internal processes in climatic variation. (National Academy of Sciences.)

theories of climatic change, although we probably could turn up at least a dozen or two without much difficulty, but will confine our discussion to several that appear most plausible in terms of what we now know about climate.

When climatic change is discussed, a term that is often used without being very precisely defined is *climate system*. We will start out by defining the term in the way we will be using it in our discussion of climate theory. When we refer to the climate system (see Figure 4), we will mean three interacting terrestrial components: the atmosphere, the hydrosphere, and the cryosphere—our planet's envelopes of air, water, and ice. Although two other elements, the lithosphere and the biosphere, are often considered internal to the climate system, we are not including them because they do not interact with the atmosphere, hydrosphere, and cryosphere in the same ways that those components interact with each other.

The state of the climate at any given time is determined by interactions between these internal components of the system as well as by some external forces. These external "boundary conditions," to use the language of the climate theorists, include the state of the sun, which supplies the energy that drives the system; the composition of the atmosphere through which this radiant energy must pass to get into the system; and the surface configuration of the lithosphere—the mass of rock and soil that underlies the atmosphere, hydrosphere, and cryosphere. The climate system is also affected by variations in the earth's orbit relative to the sun and by some other properties of the lithosphere besides the shape of its surface. The biosphere—the plants and animals that live on the earth's surface and in its air and water—can also influence the climate system.

What do we mean when we refer to the state of the climate? The National Academy of Sciences (1975) report *Understanding Climatic Change* defines climatic state as "the average (together with the variability and other statistics) of the complete set of atmospheric, hydrospheric, and cryospheric variables over a specified period of time in a specified domain of the earth-atmosphere system."

What are the variables to which this definition refers? The same report classifies these fundamental properties of the climate system:

Thermal properties, which include the temperature of the air, water, ice, and land.

Kinetic properties, which include the wind and ocean currents, together with the associated vertical motions, and the motion of ice masses.

Aqueous properties, which include the air's moisture or humidity, the cloudiness and cloud water content, ground water, lake levels, and the water content of snow and of land and sea ice.

Static properties, which include the pressure and density of the atmosphere and ocean, the composition of the (dry) air, the oceanic salinity, and the geometric boundaries and physical constants of the system.

Of the three internal terrestrial components of the climate system, the atmosphere is the most variable and has the shortest response time to external influences. It would take the atmosphere about a month to adjust itself, by transferring heat vertically and horizontally, to a temperature distribution imposed by external forces, such as a fluctuation in the input of energy from the sun. A month is also the estimated time that it would take the atmosphere to run down—for its kinetic energy to be dissipated by friction—if its external energy source, the sun, were somehow shut off.

The hydrosphere—the oceans and other bodies of liquid water—is tied very closely to the atmosphere but has a much slower response time and thus acts as a brake in some ocean–atmosphere interactions. The upper layers of the ocean interact with the atmosphere on time scales of months to years, but temperature and circulation changes in the ocean depths may take centuries.

Changes in the cryosphere—the earth's mantle of ice and snow—occur on two different time scales. Sea ice and snow cover on the land change mainly with the seasons; large areas of the temperate zones are covered with snow and ice in the winter and free of snow and ice cover in the summer. Glaciers and ice sheets respond much more slowly, as we discussed in Chapter 3, integrating long-term temperature changes over periods ranging from decades to many millions of years. Snow and ice cover also affects the earth's temperature by changing the planet's albedo—the relative amounts of incoming radiation that are absorbed and reflected by the earth's surface. Much more solar energy is reflected back into space by snow and ice than by rock or bare earth, and it seems clear that if snow and ice cover increased beyond some critical extent, so much energy would be reflected that a runaway ice age would result, producing irreversible global glaciation. The probability that this process could be triggered by any foreseeable cause appears to be very low, however.

The lithosphere, which includes mountains, ocean basins, surface rocks, sediments, and soils, changes on a much longer time scale than any of the internal components of the climate system. Mountain building, continental drift, sea-floor spreading, and other major changes in the lithosphere occur over tens and hundreds of millions of years. But they may have been a major factor in the onset and conclusion of prehistoric glacial periods through their influence on atmospheric and oceanic circulation and on the patterns of absorption and radiation of energy by the surface of the earth.

The biosphere may influence climate fairly directly on short time scales and over comparatively small areas. Several theories of desertification are based on the influence of reduced plant cover on albedo and vertical air motions. The biosphere includes the human species, and it seems clear now that, although human activities probably played a minor role in past climatic fluctuations, they almost certainly have the potential for influencing the future course of the climate.

The term *climatic change* has been used a great deal in recent years, by scientists as well as journalists and others outside the scientific community, without being defined in any precise way. All sorts of climatic variations, ranging from the onset of glacial epochs to one-year droughts in the corn and wheat fields of North America, have been lumped together as climatic changes.

One climatologist, Helmut Landsberg of the University of Maryland, has proposed a simple scheme for classifying climatic variations that is logical in terms of both time scale and magnitude. He defines a climatic *change* as "a long-lasting shift of various elements such as temperature and rainfall to new equilibrium positions." Such climatic changes, according to Landsberg, take place over thousands of years and may last tens of thousands. Variations that occur over periods of centuries, decades, or a few years are defined by Landsberg as climatic *fluctuations*. According to this system of classification, the most recent climatic change was the withdrawal of the continental ice sheets in the Northern Hemisphere 10,000 to 20,000 years ago. Climatic fluctuations that have occurred since then include the Little Ice Age from the mid-1500s until the mid-1800s and the North American Dust Bowl drought of the 1930s. Without debating the question of whether or not Landsberg's categorization is valid in any absolute sense, we will accept it for purposes of our discussion of theories of climatic variation.

The possibility exists that climatic changes—the onset of glacial epochs or the ebb and flow of continental glaciers within those epochs—occur in a completely random and unpredictable way. But many atmospheric scientists believe that climatic change is governed by physical mechanisms that, at least in theory, should be susceptible to human analysis and understanding and probably to some degree of prediction.

One definition of climate, designated in *Understanding Climatic Change* as the "physical definition," is: "the equilibrium statistical state reached by the elements of the atmosphere, hydrosphere, and cryosphere under a set of given and fixed external boundary conditions." In other words, according to this definition, if one of the boundary conditions, such as the input of energy from the sun or the configuration of the earth's surface, changed significantly, the atmosphere, hydrosphere, and cryosphere would respond to the change with a series of interactions that would eventually settle down to a new climatic state—a glacial period replacing an interglacial, for example.

But this definition, so neat and tidy in the abstract, may not be very useful in considering the behavior of the real climate system. For example, complicated and extensive sequences of responses to a change in a single boundary condition might take so long to work themselves out that the boundary conditions could change again before the climate system had settled down to a new equilibrium state. Furthermore, more than one climatic state may be able to exist under a given set of boundary conditions.

The possibility of several possible climatic states for a single set of boundary conditions, first raised by Edward Lorenz of the Massachusetts Institute of Technology, is based on the concept of transitive and intransitive systems. A *transitive* climate system would respond to a change in some critical boundary condition in the way that we described above: Regardless of its initial state, it would shift to a new equilibrium state dictated by the new set of conditions. An intransitive system would not change; it would ignore the new boundary conditions and remain in the same state regardless of shifts in boundary conditions.

The climate system clearly is not completely intransitive, and most climate theoreticians would like to think that it is transitive; that theoretically it should be possible to understand what climatic state will be dictated by a given set of boundary conditions. But Lorenz has suggested that the climate system may be a "quasi-intransitive" or "almost-intransitive" system. That is, when the boundary conditions changed, the system might appear to be stubbornly intransitive at first. But when some factor in the system reached a certain level, the system would shift to another state, in which it would remain ostensibly stable. However, if some other key element then reached a certain level, the system might shift again, perhaps to the initial state or perhaps to a third state. The glacial-interglacial oscillations of the past million years could conceivably have been the transitions between two states of an almost-intransitive system.

The question of the predictability of climatic change hinges on this question of transitivity to a great extent. If the climate system is transitive, then it should be possible to understand its interactions and eventually to predict their outcome. If those interactions are influenced substantially by a boundary condition such as fluctuations in the output of the sun, then of course it would be necessary to be able to predict the solar fluctuations in order to predict climatic changes. An almost-intransitive climate system would present formidable problems of prediction. It might also present a great danger in terms of human impact on climate. If a human activity, such as the injection of massive quantities of carbon dioxide and waste heat into the climate system from energy production, turned out to be the influence that triggered the shift of an almost-intransitive climate system from one state to another, then reducing the magnitude of the influence would be unlikely to reverse the process.

Even if the climate system is transitive, it clearly includes a good many closed subsystems in which two or more factors mutually influence each other. It has been suggested that volcanic activity might influence glaciation and that glaciation, by changing patterns of pressure on the earth's crust, might be a factor in triggering volcanism. On a more modest scale, the wind can drive an ocean current that alters the distribution of sea-surface temperatures, which in turn can influence the large-scale atmospheric circulation that determines the strength and direction of the wind. The chain of causation is essentially circular; it would be arbitrary to label one element as the cause and another as the effect.

Such circular sets of interactions can comprise either positive or negative feedback loops. For positive feedback, the elements reinforce one another, so that the circular chain of interactions continues to increase in strength. Negative feedback involves a net weakening of the interactions, so that they tend to damp themselves out. Although the climate system almost certainly contains both kinds of feedback loops, there is considerable uncertainty about which parts of the system may be self-regulating and which ones may have the potential for runaway reactions to certain triggering mechanisms, some of which might be set off by human activities.

Some theories of climatic change and fluctuation are based primarily on external causes—changes in the boundary conditions—while others invoke internal mechanisms. Still other theories combine these two approaches. The climatic changes that have received the most attention from the theorists have been the Pleistocene ice ages of the last one million years—the glacial periods that "periodically bury the sites of Boston, Birmingham and Berlin under great thicknesses of unmelted snow," as the British science writer Nigel Calder alliteratively put it in a recent article. (He was referring to Birmingham, England, of course; there is no evidence that the North American ice sheet came anywhere near the state of Alabama even during the most severe glacial episodes.)

On a longer scale, perhaps the largest question of climatic change concerns the causes of the anomalously cold epochs, such as the Pleistocene, that have interrupted Brooks' warmer and more uniform normal climate of geological time. On shorter time scales, climatologists have been concerned with anomalous periods of a century or two, such as the Little Ice Age, as well as with even briefer anomalies within centuries or decades.

One of the earliest theories of glaciation was proposed by Jens Esmark of Norway in the 1820s. He suggested that the earth's elliptical orbit around the sun had varied substantially enough to cause the great climatic swings between glacial and interglacial periods. Although the orbital variations that Esmark proposed were not real, later theorists, including the Scottish climatologist James Croll, the English astronomer Robert Ball, and the German geophysicist Alfred Wegener, proposed hypotheses to account for glaciation on the basis of fluctuations in the earth's orbit.

Such theories of climatic change have always seemed plausible, if only because the seasons themselves, which represent substantial fluctuations in climate, are based on changes in the earth's position relative to the sun that repeat themselves on a regular annual cycle. Astronomers had established the reality of other, longer term variations in the earth's attitude and orbit. The eccentricity of the earth's orbit changes; the planet rolls, changing the tilt of its axis relative to the path of the incoming solar radiation; and the "wobble" known as the precession of the equinoxes causes a slight variation in the time of year that the earth approaches the sun most closely, so that the relative lengths of summer and winter change very slightly from year to year.

Building on Wegener's work, Milutin Milankovitch of Yugoslavia developed the hypothesis that usually bears his name. Milankovitch postulated that changes in the earth's position relative to the sun, which occur in calculable cycles, would produce the most severe glaciation in the Northern Hemisphere at times when the earth was not tilted on its axis and the summer occurred when the earth was farthest from the sun. The warmest interglacial periods would come when the planet was tilted toward the sun and also approached it most closely during the Northern Hemisphere summer. If the Milankovitch hypothesis is valid, the precession of the equinoxes would produce a 20,000-year cycle; the changing tilt of the earth's axis would cause a 40,000-year cycle; and changes in the eccentricity of the earth's orbit would result in a 90,000- to 100,000-year cycle of climatic change.

However, after achieving some prominence in the 1930s, the Milankovitch hypothesis fell into disrepute, mainly because the astronomical cycles on which it was based did not match the paleoclimatological chronology of glaciation that was generally accepted at that time. Until the mid-1950s, it was assumed that there had been four or perhaps five periods of major glaciation during the last million years or so—the long Pleistocene cold spell. But starting around 1955, as the archives of ocean-bed core samples grew and new radioisotope and ecological techniques for analyzing them were developed, a very different picture of glacial paleoclimatology began to emerge.

Cesare Emiliani, now of the University of Miami, was one of the pioneers in this new reconstruction of ice-age chronology. Using techniques that had come out of nuclear energy research during World War II, he developed the oxygen-18 correlation to glacial conditions that we described briefly in Chapter 3. His first analyses indicated that there had been seven periods of severe cold instead of the previously accepted four ice ages, and that the interglacials between them were considerably shorter than the 100,000 to 250,000 years that had been postulated in the four-ice-age chronology.

In the early 1960s, George Kukla of Czechoslovakia (now with the Lamont-Doherty Geophysical Observatory of Columbia University) analyzed deposits of soil in his country and concluded that there had been at least ten glacial periods. But it was difficult to correlate Kukla's soil deposits and Emiliani's ocean-bed cores chronologically. Such correlation was done in the late 1960s, when the discovery of reversals in the earth's magnetic field provided a new dating tool. Neil Opdyke and his colleagues at Lamont-Doherty found evidence of the last major reversal of the earth's north and south magnetic poles, which occurred about 700,000 years ago, at the point in the ocean-bed cores that corresponded to the eighth ice age before the present. The same event was detected in the Czechoslovakian soil samples at the level where their eighth ice age had been detected.

The beginning of the CLIMAP project in the early 1970s brought together the efforts of many of the scientists who had been working on ice-age chronology. Nicholas Shackleton of Cambridge University refined the technique of oxygen-18 dating. James Hays of Lamont-Doherty and John Imbrie of Brown University improved ecological methods of reconstructing climate by analyzing the forams and other organisms in ocean-bed cores. Hays, Imbrie, and Shackleton compared the new ice-age chronology, which indicated that there had been as many as 20 periods of major glaciation, with Milankovitch's astronomical cycles. They found stronger and stronger evidence of correlations between the two sets of fluctuations. Finally, they felt certain enough to present their case to the climatological community. First in a paper presented at an international conference on climatic change held in Norwich, England, in August 1975, then in an article in *Science* magazine in December 1976, Hays, Imbrie, and Shackleton concluded that "changes in the earth's orbital geometry are the fundamental cause of the succession of Quaternary ice ages." They cited three distinct peaks in the graph of climatic fluctuation, at periods of 23,000, 42,000, and approximately 100,000 years. These correspond closely to the dominant periods of the precession of the equinoxes, the tilt of the earth's axis, and the eccentricity of the earth's orbit. Although the mechanisms by which these astronomical variations produce the ice ages are not understood, the circumstantial evidence for cause and effect is very strong.

Does this mean that the nine and sixty theories of climatic change have now been reduced to one? The answer is emphatically *no*. Although the CLIMAP researchers appear to have examined their evidence rigorously and to have tested it carefully, their vindication of Milankovitch deals with climatic change on one time scale: that of the ebb and flow of the ice sheets within the Pleistocene. It does not tell us what triggers the beginning of such frigid interruptions of the bland normal climate of geological time. Nor does it say anything about the droughts and floods and freezes that represent the most immediate climatic threats to food production and other essential human activities.

The question of how cold epochs such as the Pleistocene begin is not answerable in terms of the evidence and conclusions of Hays, Imbrie, and Shackleton. As Cesare Emiliani put it several years ago, "The main difficulty with the Milankovitch theory is that it fails to explain why the Ice Epoch developed only recently—within the last million years—after 200 million years during which there was no ice."

To account for the onset of the Pleistocene Epoch, Emiliani returned to the geographical influences that C. E. P. Brooks had espoused with such fervor. During the period between the two world wars, Brooks published a series of books in which he put forward and developed the theory that the glacial epochs were the result of periodic "revolutions" in the earth's crust every 250 million years. During these periods, which lasted about 50 million years, mountains rose from land that had been flat and comparatively featureless. Through a number of mechanisms, including disruption of the global atmospheric circulation by mountain ranges and reductions in the extent of the oceans caused by changes in the ocean beds that accompanied the upward thrust of the mountains, these geological revolutions must have affected the climate in substantial ways. By correlating the geological chronology of mountain building with the evidence of the onset of glacial epochs, Brooks built a detailed case for geographical influences as the principal determinant of glaciation. However, the case was not very convincing on the scale of glacial and interglacial periods within the epochs, as mountain ranges clearly had not popped up and down every few hundred thousand years.

Emiliani accepted the Milankovitch hypothesis as the explanation for these shorter term fluctuations, then invoked geographical changes to account for the cold epochs. His analyses of ocean-bed cores indicated that the average global temperature had reached a maximum about 85 million years ago, at a time when there were extensive shallow seas and only a few low chains of mountains. Then a geological revolution began. Huge mountain ranges—the Alps, Andes, Himalayas, and Rockies—began to rise, and a large part of the Pacific Ocean floor began to sink. Millions of square miles of land that had been under water were exposed. Since dry land absorbs less solar radiation than water does, the planet grew cooler. Polar ice caps began to grow, and the ice absorbed even less solar energy. Emiliani theorized that eventually this process had cooled the earth off sufficiently that one of the cool-summer cycles of the Milankovitch hypothesis could have caused a great increase in Northern Hemisphere ice, reducing the absorption of solar radiation even more. At that point, Emiliani believed, the ice epoch was well under way.

But as the oceans grew colder, on a much longer time scale than that of the cooling of the atmosphere, the mechanism fed back on itself. Cooler oceans mean less evaporation, less atmospheric water vapor, and less

snowfall. Eventually the winter snow accumulation was less than the reduction of snow and ice by summer melting. A series of warm summers,
resulting from a new phase of the Milankovitch cycle, would then reduce
the ice cover. However, until the great mountain chains have eroded
away, the ice will spread again at the appropriate points in the cycle of
wobble, orbital eccentricity, and precession of the equinoxes. According
to this theory, as long as the Alps, Andes, Himalayas, and Rockies thrust
upward into the atmosphere, we will remain in a glacial epoch, and the
ice will advance and retreat with the earth's position relative to the sun.

Emiliani's explanation, although not conclusively substantiated by any
means, is interesting for several reasons. First, it heeds Brooks's admonition that the greatest extremes of climate are not necessarily caused by
one factor but by a combination of several influences pushing in the same
direction. Second, it involves both external and internal features of the
climate system. The change is triggered by one of the boundary conditions—the configuration of the earth's crust—but develops on a course
that is strongly influenced by interactions of the atmosphere, hydrosphere, and cryosphere.

William Donn and Maurice Ewing of the Lamont-Doherty Geological
Observatory have put forward a theory of glaciation that resembles
Emiliani's in that it proposes a combination of changed boundary
conditions and internal interactions as the explanation for the Pleistocene
ice ages. (However, Emiliani and Donn have criticized each other's work
severely in the pages of scientific journals, with each maintaining that the
other's theory is completely wrong.)

The Ewing-Donn theory is based on the hypothesis, which is supported
by extensive but not completely conclusive evidence, that the North and
South Poles of the earth have "wandered." The theory of polar wandering
grew out of the theory of continental drift, which has become generally
accepted in recent years. The scientist who did most of the early work on
the continental drift theory was the German geophysicist Alfred
Wegener, who also built much of the foundation on which the
Milankovitch hypothesis rests. Very simply, the continental drift theory
proposes that the present continents, or at least those in the Southern
Hemisphere, were once joined together, but they drifted apart to form
South America, Africa, Australia, Antarctica, and India. (Wegener
originally proposed that all of the present continents were joined in a
great land mass that he called Pangaea, but contemporary geophysicists
doubt that North America, Europe, and northern Asia were ever joined
with the Southern Hemisphere continents.)

Acceptance of the continental-drift theory implies that the earth's
crust is a thin shell of solid material floating on a layer of highly viscous

rock that nevertheless is enough of a fluid to permit the crust to move horizontally. The polar-wandering theory, based on fossil evidence as well as geological clues to long-past changes in the earth's magnetic field, proposes that, in addition to movement of the continents relative to one another, the entire crust of the planet, including the ocean beds as well as the land masses, has shifted relative to the earth's core, whose axis of rotation determines the location of the poles.

The Ewing-Donn theory, which has a good deal of scientific elegance, proposes that polar wandering is the primary cause of both the onset of the present glacial epoch and the succession of glacial and interglacial intervals within that epoch. Although this theory has been revised in some of its particulars since it was first published in *Science* in 1956, its essential elements have remained simple and constant. The start of a glacial epoch, it maintains, is caused by a polar shift that puts the North Pole about in the middle of the landlocked Arctic Ocean, and the ensuing sequence of ice ages and interglacials results from a series of interactions between evaporation, precipitation, and glaciation that recur within the internal elements of the climate system.

The sequence of events that causes the present glacial epoch began, according to Ewing and Donn, when the poles migrated to their present position about 35 million years ago. The North Pole moved from a location in the northern Pacific Ocean, which was influenced by warm currents flowing from the south, to the middle of the Arctic Ocean, which is almost completely surrounded by land masses that block any large exchanges of water with the other oceans. At this point the Arctic Ocean was free of ice, and a sufficient supply of moisture could evaporate from it to feed storm systems that deposited snow over northern Canada, Greenland, and Siberia. Glaciers began to form, in Antarctica as well as in these northern regions, and their high reflectivity cooled the earth more. By the time the Arctic Ocean froze over, cutting off the source of moisture that had fed the glaciers in their early stages, the ice sheets had reached far enough south over Europe and North America so that they were fed by moisture from the North Atlantic and the Gulf of Mexico. The Siberian ice, cut off from southern moisture by the Himalayas, stopped growing when the Arctic Ocean became ice covered.

Finally, the North American and European ice sheets reached their southernmost extent, where summer melting removed as much ice as was added by winter snowfall. Cool air from the ice sheets and icebergs that broke away where the glaciers reached the shore were cooling the North Atlantic, and as the ocean grew cooler, less moisture evaporated from its surface into the atmosphere. Less atmospheric moisture meant less snow, and as the winter build-up of snow grew less than the summer melt-off, the glaciers retreated northward. With the Arctic Ocean frozen, there was no moisture source available in the north polar region, so the ice continued to retreat. As the ice cover grew smaller, more solar energy was absorbed by the earth, and temperatures began to rise. This, according to Ewing and Donn, melted the ice cover on the Arctic Ocean, paving the way for the whole cycle to start over. And this cycle, repeated over and over, accounts for each of the glacial-interglacial successions within the Pleistocene Epoch.

The Ewing-Donn theory is ingenious and plausible in many respects, but it ignores a number of factors, such as mountain building, that almost certainly play some role in climatic change. Although an ice-free Arctic Ocean probably would affect the climate system in some significant ways, the Ewing-Donn theory seems to many climate theorists to present too neat and tidy an explanation for the Pleistocene fluctuations of the polar ice. However, some scientists have expressed concern that southward diversions of some Siberian rivers for irrigation, which have been proposed by the Soviet Union, might have widespread impacts on the climate by changing the salinity of the Arctic Ocean, creating much larger ice-free areas than now exist, and possibly triggering sequences of climatic interactions such as those proposed by Ewing and Donn.

Another theory of climatic change is based on the effects on the transparency of the atmosphere of two of its constituents: volcanic dust and carbon dioxide. According to this hypothesis, dust in the atmosphere blocks off incoming short-wave solar radiation but allows the long-wave heat energy radiated by the land and the oceans to pass freely back out into space. Carbon dioxide, on the other hand, allows the short-wave solar radiation to enter the atmosphere easily but holds back the heat energy radiated upward from the earth's surface. Thus, all other things being equal, an increase in dust in the atmosphere should have a cooling effect, and an increase in atmospheric carbon dioxide should cause a warming trend.

The hypothesis that volcanic dust produces cooler temperatures, at least on a time scale of a few years, it supported by a good deal of circumstantial evidence. Benjamin Franklin suggested in 1784 that the severe winter that year was the consequence of volcanic eruptions in Iceland and Japan the previous year. The terrible winter of 1816—"the year without a summer"—followed three major volcanic eruptions: Soufriere on St. Vincent Island in 1812; Mayon in the Philippines in 1814; and Tambora in Indonesia in 1815.

Quantitative measurements made at Montpellier Observatory in France after the eruption of Krakatoa in 1883 supported the theory. This eruption in Indonesia threw 13 cubic miles of debris into the atmosphere.

FIGURE 5   Volcanic eruptions, such as this eruption of Alaska's Mount Augustine in 1977, can throw enough ash and other particles into the atmosphere to affect the input of solar energy into the climate system. (National Center for Atmospheric Research.)

Volcanic ash at altitudes up to 50 miles blocked off the sun, bringing 57 hours of darkness 50 miles from Krakatoa and 22 hours without sunlight at a distance of 130 miles. As the dust drifted around the world, it produced vivid sunsets and strange haloes around the sun and moon. When the dust reached Europe several weeks after the eruption, scientists at the Montpellier Observatory observed a 20 percent drop in direct solar radiation. For about three years, until the dust dissipated, the solar radiation at Montpellier remained about 10 percent lower than normal.

In 1913, William Humphreys, a U.S. Weather Bureau scientist, published a paper in which he documented the historical correlation between volcanic eruptions and cold periods and presented the evidence for a cause-and-effect relationship between the two. Harry Wexler, who headed the Weather Bureau half a century later, was a strong supporter of this theory, and contemporary climatologists such as Hubert Lamb also believe that there is a definite relationship between volcanic eruptions and cool periods. Aircraft have been used to sample the particles thrown into the atmosphere by recent eruptions (Figures 5 and 6).

However, the case for volcanic activity as the primary cause of glacial

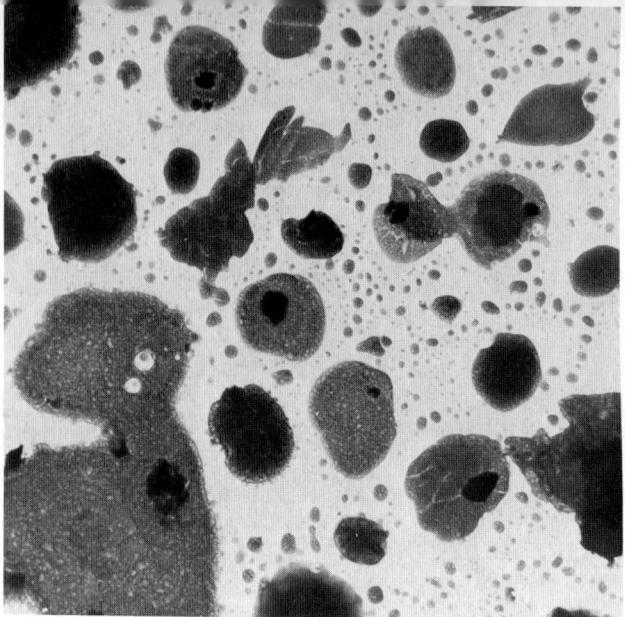

FIGURE 6   This electron micrograph of volcanic-fume samples from the Mount Augustine eruption in 1977 shows a large number of sulfuric-acid particles, identifiable by the configuration of small droplets surrounding a larger central droplet. Sulfuric acid is only one of many constituents of the plume thrown high into the atmosphere when a volcano erupts. (Richard Cadle, National Center for Atmospheric Research.)

periods is not very solid. James Kennett and Robert Thunell of the University of Rhode Island recently attempted to correlate volcanic ash in ocean-bed cores with glacial periods. They reported in a 1975 *Science* article that "greatly increased volcanism during the last two million years closely coincides with that interval of earth history marked by major and rapidly oscillating climatic conditions related to glacial-interglacial cycles in the Northern Hemisphere." But they pointed out that the data were not fine enough to discern any cause-and-effect relationships between volcanic eruptions and increased glaciation within this period. They also referred to a suggestion by R. K. Matthews of Brown University that increased volcanism might sometimes be an effect of glaciation, instead of a cause, produced by geological stresses resulting from the shifting of great masses of water between the oceans and continental ice sheets. This hypothesis suggests that volcanic eruptions might be considered an interacting element in the internal atmosphere-hydrosphere-cryosphere climate system rather than a boundary condition, and points up the difficulty of sorting out the elements of climatic change and fluctuation in any neat and systematic way.

The effect of carbon dioxide on global temperature has traditionally been known as the greenhouse effect, on the presumption that the atmospheric carbon dioxide plays the same role as the glass roof of a greenhouse. However, some atmospheric scientists have pointed out that the warming effect of the greenhouse roof results more from trapping warm air inside the structure than from its property of admitting short-wave and retarding long-wave radiation. At any rate, greenhouse effect or not, an increase in atmospheric carbon dioxide would be expected to cause a warming trend, all other things being equal (see Figure 7).

For many years, increasing atmospheric carbon dioxide was accepted by many scientists as the cause of the long warming that occurred between the late 1800s and the 1940s. The beginning of this warming

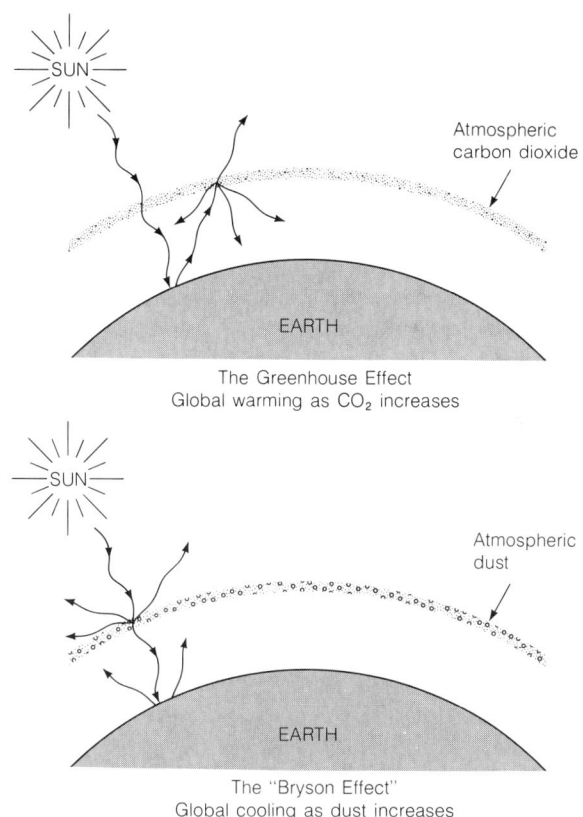

FIGURE 7 Some scientists believe that the fluctuating transparency of the atmosphere is the key factor in climatic variations. For example, carbon dioxide tends to produce a net warming effect, while certain kinds of particles might be expected to produce a net cooling, as indicated in these simplified diagrams. (Henry Lansford.)

coincided with the Industrial Revolution, and steadily increasing amounts of fossil fuels—coal, oil, and gas—were burned by industry during that period. By the 1940s, many scientists had accepted the theory that the so-called greenhouse effect was responsible for the warming trend and that this trend would continue as long as industrial combustion of fossil fuels continued to increase.

But starting around 1950, the Northern Hemisphere mean temperature began to creep downward. This trend resulted in another hypothesis, based on what Reid Bryson of the University of Wisconsin likes to call the "human volcano." According to Bryson, the warming trend of the first part of the twentieth century was indeed caused by increasing carbon dioxide levels in the atmosphere. But around the middle of the century, this effect was overwhelmed by an increase in particles thrown up into the atmosphere by human activities, primarily agriculture. Increasing cultivation of semiarid lands, along with slash-and-burn agriculture in which large areas of tropical jungle are burned off periodically for farmlands, contributed large quantities of dust and smoke particles to the atmosphere, intercepting part of the solar radiation before it could enter the climate system. Bryson called the agricultural dust and smoke and the particulate matter that is thrown up into the atmosphere from industry and other human sources the human volcano. He maintains that human activities are sending dust into the atmosphere at a greater rate than volcanic eruptions that have caused cooling trends in the past.

Bryson also believes that the cooling, which has been more marked at higher altitudes, has changed shorter term climatic patterns by altering the global atmospheric circulation. Briefly, he proposes that the cooling temperatures in northern regions have made the jet stream and other high-level, globe-circling winds behave in the summer more as they do in the winter—that is, remaining in a southerly location and exhibiting more vigor. The result, Bryson believes, has been more variable summer weather—an increase in floods, droughts, and other extreme weather events. He also believes that the Sahelian drought occurred because the monsoon over West Africa was suppressed by this southerly displacement of the upper-level westerlies.

Bryson's theory of global warming and cooling is not generally accepted by atmospheric scientists, both because it seems to many to be impossible to verify and because there are indications that the cooling trend has halted and that a new warming trend may be underway. Bryson's critics also point out that different kinds of particles at different levels of the atmosphere may have diverse effects on temperature; under some conditions, an increase in particulate matter might cause warmer rather than cooler temperatures. Moreover, the phrase that we used in introducing the idea that dust might cause cooling and carbon dioxide, warming—all other things being equal—is very important. In the climate system, it is highly unlikely that any one factor ever changes without others changing at the same time, either through interactions or simply by coincidence. The behavior of the atmosphere alone is fiendishly complex, and the fact that it is coupled to the other elements of the climate system means that no single process should be considered in isolation from all the others that are going on at the same time.

There is one boundary condition of the climate system that, if it did vary significantly, could be expected to cause sizable climatic fluctuations and changes: the input of solar energy that drives the system. But the traditional view of the sun has been that it is constant in its total energy output. The sun, our nearest star, has always stood out as an apparently stable and unvarying source of light and energy among the 500 billion or so mostly variable stars that compose our galaxy. The prevailing view among solar physicists has been that the total radiation that reaches the top of the earth's atmosphere from the sun does not vary over time by more than a fraction of a percent; this input of energy has traditionally been referred to as the solar constant.

Over the years, a few scientists have taken issue with the concept of the solar constant. In the mid-1950s, Ernst Öpik of Armagh Observatory in Ireland proposed that past changes in the internal processes of the sun had influenced its energy output and caused climatic changes on earth. Stated simply, his theory was that certain elements in the sun—magnesium, carbon, iron, silicon, oxygen, neon, and nitrogen—which he lumped together as "metals," accumulate periodically around the core of the sun, resulting in a spasmodic expansion of the sun that absorbs energy, resulting in a reduction in the amount of energy radiated into space. This periodic solar convulsion, Öpik believed, triggered past ice ages on earth. He also theorized that solar changes over the next 1 billion years would cause the earth to grow steadily hotter. However, most scientists did not accept the physicial assumptions about the sun on which Öpik's theory rested, and the concept of the solar constant continued to prevail.

An important characteristic of the sun that has long been considered roughly but reliably cyclical is the rise and fall of sunspot activity. Sunspots are dark blotches on the face of the sun. Galileo is usually credited with discovering them, but John Eddy, a solar astronomer with a historical bent, has pointed out that at least three other European scientists observed sunspots with telescopes at about the same time that Galileo saw them in the early 1600s. Because the sun was considered an example of the perfect handiwork of God, the notion of blemishes on its face was at first resisted as heretical, but their existence eventually was accepted.

By the mid-nineteenth century, astronomers had discovered that the number of sunspots rose to a maximum and fell to a minimum on a cycle that averaged 11 years, although it sometimes was as short as 8 years or as long as 15. This fairly regular solar cycle came to be accepted as a characteristic of the sun that had probably always prevailed and always would.

It was not until the early twentieth century that science solved the riddle of what sunspots really are. Although theories that sunspots might be clouds over the sun or solid objects between the earth and the sun had been proposed and discarded, nobody had really explained their nature satisfactorily. In 1908, the American astronomer George Ellery Hale developed an instrument that could measure magnetic fields on the sun. He used this magnetograph to determine that sunspots are in fact giant magnetic fields 1000 times as strong as the earth's magnetic field, covering areas on the sun that are larger than the earth. These magnetic fields, created by motions of the electrically charged particles that compose the gaseous atmosphere of the sun, produce the dark areas that we see as sunspots by blocking the flow of hot, luminous gas from the interior to the surface of the sun (see Figure 8). Other manifestations of solar activity—solar flares, prominences, and coronal streamers—are also shaped by the magnetic fields that produce sunspots.

In recent years, the magnetograph has been used to establish that the large-scale magnetic field of the sun reverses itself each time the number of sunspots reaches a maximum, so that at one sunspot minimum the sun's magnetic field is parallel to the earth's magnetic field and at the next one it is opposite to it. Thus there is a 22-year double sunspot cycle superimposed on the 11-year one.

The solar corona—the outermost part of the sun's atmosphere—is the bright, irregular halo that is visible around the black disc of the moon during a total solar eclipse. The high temperatures of the corona drive the solar wind, made up of electrically charged subatomic particles—protons and electrons—that flow continuously out into space. When the solar wind reaches the earth's magnetic field, it deforms it, pulling it out into a sort of teardrop shape extending away from the sun. As the particles of the solar wind collide with nitrogen molecules and oxygen atoms in the earth's upper atmosphere, they take electrons from them, leaving them ionized, or electrically charged. Storms on the sun—violent disturbances in solar magnetic activity—produce disturbances in the solar wind that result in electrical activity in the earth's upper atmosphere. This manifests itself in brilliant displays of the aurora borealis—the northern

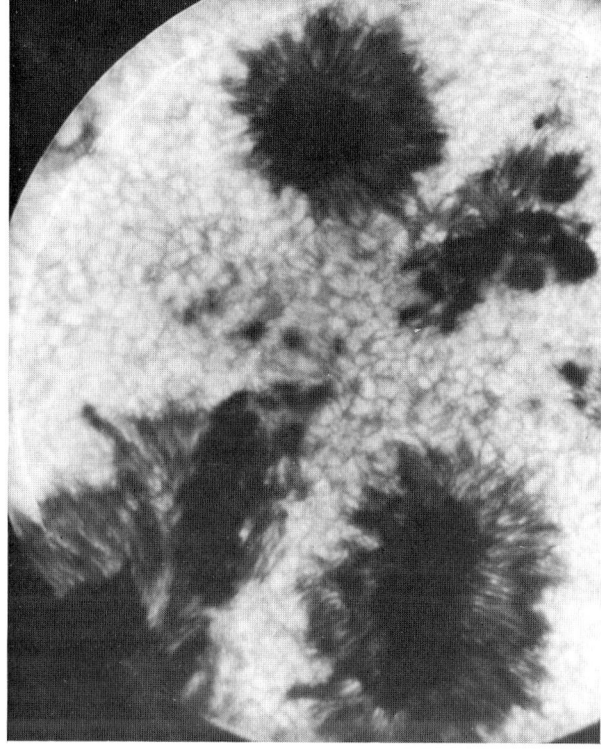

FIGURE 8 This "close-up" view of several sunspots shows their structure. The dark region in the middle is called the umbra, and the fibrils extending outward constitute the penumbra. (Sacramento Peak Observatory.)

lights—and disturbances in the earth's magnetic field known as geomagnetic storms, which can affect radio communications, long-distance telephone calls, electric power transmission, and electronic navigation systems. However, the amounts of energy involved in these phenomena of the upper atmosphere have always been considered far too small to have any direct effect on weather systems. And there did not appear to be any connection between fluctuations in sunspot activity and the total output of solar radiation that drives the earth's weather and climate machine.

So much for the conventional view of solar-terrestrial relationships. As with our earlier discussion of weather, we have dealt cursorily with a subject that is complex and fascinating in its own right, and we encourage the reader who is so inclined to pursue it further. We shall move on now to some theories that run counter to the ideas about solar influences on the earth's weather and climate that have prevailed generally.

After it was discovered that sunspots were in fact blotches on the face of the sun, it seemed reasonable to some scientists to assume that they might block off part of the sun's energy output, and they speculated that such reductions in energy might cause climatic variations. As early as 1802, one astronomer suggested that the rainy regions of the tropics showed temperature fluctuations that varied inversely with the number of sunspots—many sunspots accompanied cooler temperatures. This seemed quite reasonable and logical at the time, as the sunspots were dark and presumably lowered the sun's radiant output.

Later researchers worked to relate patterns of weather and climate to fluctuations in the activity of the sun, but the results were not clear-cut, and, for the most part, the scientific community remained skeptical. About the time of World War II, some careful scientific work that was done in Germany suggested that barometric pressures over northern Europe changed systematically following large outbreaks of sunspots and solar flares and the subsequent auroral displays and geomagnetic disturbances, which were accepted as consequences of solar activity, particularly outbreaks of solar flares.

A decade or so later, several American scientists gave serious attention to this subject. One was Ralph Shapiro, of the Air Force Cambridge Research Laboratories, who developed a new measure of the stability of North American weather systems, which he called the persistence correlation index. When the weather remained relatively unchanged from one three-day period to the next, this index had a high value. When the weather over the grid of North American weather stations was changing, the index dropped to a lower value. Shaprio showed that the average behavior of the persistence correlation index seemed to be significantly related to the occurrence of large geomagnetic disturbances, which were taken as evidence of high solar activity. For several days immediately following the geomagnetic disturbances, the weather was stable and the index was high. Then the persistence correlation index would steadily decline over the next week or so.

It was about this time that I became deeply interested in this subject. (The I in this discussion is Roberts, the scientist of our scientist-writer team, who has been studying certain aspects of solar-terrestrial relations for many years.) Although my main interest was in possible linkages between solar activity and terrestrial climate, I felt that solid evidence of short-term relationships between solar activity and weather would establish a foundation for possible solar-climate relationships. By the late 1950s, a few colleagues and I had become convinced that low-pressure cyclonic weather systems that developed over the Gulf of Alaska following great auroral displays and geomagnetic storms grew larger and more intense than such systems that were not preceded by auroral and geomagnetic events.

More recently, John Wilcox of Stanford University and some other scientists have continued and improved these analyses, extending them to more of the Northern Hemisphere. The magnetic field between the sun and the earth is divided into sectors of alternating polarity. Wilcox and his colleagues have demonstrated convincingly that there are changes in the behavior of strong cyclonic weather systems in the Northern Hemisphere after the earth passes through one of these interplanetary magnetic-field sector boundaries, which are related to the distribution of magnetic fields and sunspots on the sun. Two scientists at the University of Toronto, Colin Hines and Itamar Halevy, set out somewhat skeptically to test the results that had come from the work at Boulder and Stanford. Through more rigorous research, they confirmed that the relationship between sector boundary crossings and the intensity of weather systems does indeed appear to be real. However, none of these researchers has developed a physical explanation for the relationship. Although the evidence is very convincing, it remains statistical, or circumstantial as it would be termed in a court of law.

In addition to this rather convincing case for a relationship between solar activity and terrestrial weather systems, there is good circumstantial evidence for a strong correlation between the sunspot cycle and a recurring climatic fluctuation in one geographical area. In Chapter 3, we mentioned the periodic droughts that have occurred on the high plains of the United States about every 20 to 22 years. There have been at least nine of these droughts in the recorded history of the region, including the one that occurred in the mid-1970s.

Scientists working in several diverse disciplines, including an astronomer, Charles Greeley Abbot of the Smithsonian Institution; a geographer, John Borchert of the University of Minnesota; and an agricultural scientist, Louis Thompson of Iowa State University, have been struck by the fact that the middle years of the high-plains droughts have coincided with alternate minima in the sunspot cycle, when the sun's magnetic field is opposite to the earth's. Although I have not been able to propose a mechanism by which the solar minimum would cause the droughts, I have always considered the correlation very provocative. The sunspot minimum of the 1970s, which arrived somewhat later than the 20-year average that has prevailed in this century, was accompanied by a corresponding lag in the onset of drought in the high plains.

One recent convert to the belief that droughts in the western United States may be correlated with the double sunspot cycle is J. Murray Mitchell, the climatologist who did much of the work that established the reality of the global warming trend of the first half of this century and the subsequent Northern Hemisphere cooling. Mitchell, who is with the Environmental Data Service of the National Oceanic and Atmospheric

Administration, said in 1975 that "the line-up of droughts into what looks like a regular cycle—whether connected with sunspot minima or not—doesn't appear to go back very far. It's a fact of only the past 100 years of drought history that we know anything about. The 'cycle' probably didn't exist before that."

However, more recently Mitchell has collaborated with Charles Stockton of the University of Arizona in a study that indicates that droughts in the western United States may be correlated with the double sunspot cycle over a much longer period of time. Stockton has analyzed tree rings from 40 sites in the West to reconstruct climate back to AD 1700. The periods of drought that show up in this record correspond fairly closely with the alternate sunspot minima. Mitchell, who has analyzed Stockton's data carefully, says that their joint conclusions support the idea that droughts in the West do indeed occur on a 22-year cycle that corresponds to the double sunspot cycle. However, a physical mechanism to establish cause and effect in this correlation still has not been identified.

The sunspot cycle has fascinated researchers for many years, and attempts have even been made to correlate it with periodic fluctuations in such unlikely statistics as the state of the stock market and the rabbit population of England. Stephen Schneider of the National Center for Atmospheric Research said not long ago that "statistical correlations in the absence of physics always lead to a fight," a statement that certainly is applicable to the apparent correlation between the sunspot cycle and high-plains droughts. Other scientists have pointed out that, by searching through enough data, one is eventually bound to find fluctuations in weather and climate somewhere in the world that correspond to solar variations. Why would the high plains in the lee of the Rockies, of all the possible regions of the world, be particularly susceptible to solar influences?

In fact, there is a unique element of the earth's surface that exercises a significant influence on the climate system in this region (Figure 9). The Rocky Mountain chain is the largest single north-south mountain barrier to the mid-latitude westerly winds that exists anywhere on earth. It may be responsible for a more or less permanent feature of the general circulation—a node or "ripple"— that causes drought on the high plains when the westerlies are strong and persistent. Perhaps this means that the "normal" climatic state of the region is drought, but that it is somehow disrupted when sunspot activity is high and the sun's magnetic field is parallel to the earth's. However, this idea is highly speculative and needs to be either proved or disproved by rigorous research.

Another apparent correlation between solar activity and terrestrial climate has been noted recently by John Eddy of the National Center for

FIGURE 9 The high plains east of the Rocky Mountains are in the lee of the largest north-south barrier to the westerly flow of the atmosphere anywhere on earth. (Henry Lansford.)

Atmospheric Research. Using evidence drawn from a variety of historical and physical sources, Eddy has established with reasonable certainty that the 11-year sunspot cycle of recent times has not always prevailed and that sunspots have virtually disappeared from the face of the sun for long periods of time. Eddy has established that prolonged periods of very low solar activity occurred from AD 1100 to 1250, from 1460 to 1550, and from 1645 to 1715. The last period is usually called the Maunder Minimum after E. W. Maunder, the British astronomer who, along with Gustav Spörer of Germany, first called attention to it. The Maunder Minimum, Eddy has pointed out, coincides very closely with the coldest part of the Little Ice Age. And a prolonged solar maximum in the thirteenth century coincides with a period when the earth's climate was generally warm. Eddy's work does not establish any physical linkages between solar activity and terrestrial temperatures, but he suggests that it may have implications for the concept of the solar constant. The mechanism involved, according to Eddy, may be small but ponderous long-term changes in the total radiative output of the sun.

But Eddy is not a fervent supporter of theories that link solar behavior with terrestrial weather and climate. He feels that the field has suffered, over the years, from a lack of objectivity. In a 1977 review in *Science* of the proceedings of a symposium on "Possible Relationships Between Solar Activity and Meteorological Phenomena" held by the National Aeronautics and Space Administration (NASA) in 1973, Eddy wrote:

> I have found that most people, scientists and otherwise, badly want there to be a connection between weather or climate and the sun. The sun surely drives the weather machine, so why aren't the variations in one linked to the vagaries of the other?
>
> The trouble, as the old-line meteorologists keep pointing out, is that weather and climate changes can happen in spite of the sun. Observable solar changes might have tropospheric effects, but not necessarily. The earth could enjoy dramatic weather and varied climate, with droughts and heat waves and ice ages, even if the sun were locked to a perfect thermostat and didn't change its output at all, in photons, particles, or whatever.

In a paper that he presented at an American Astronomical Society meeting in 1975, Eddy quoted Charles Young, whom he referred to as his favorite solar astronomer of all time. In his classic book, *The Sun,* Young had this to say about sunspots and the weather:

> In regard to this question, the astronomical world is divided into two almost hostile camps. . . . One party holds that the state of the Sun's surface is a determining factor in our terrestrial meteorology, making itself felt in our temperature, barometric pressure, rainfall, cyclones, crops, and even our financial conditions. . . . The other party contends that there is and can be no sensible influence upon the Earth produced by such slight variations in the solar heat and light. . . . It seems clear that we are not in a position yet to decide the question either way. It will take a much longer period of observation, and observations conducted with special reference to this subject of inquiry.

Pointing out that Young wrote these words in 1895 and died in 1908, Eddy said: "I will not feel especially proud to meet him, when the roll is called up yonder, to report how we have done on this important question. . . . "

Eddy found nothing really new in the proceedings of the 1973 NASA conference—"The selected statistics that excite the enthusiasts still cause the skeptics to shake their heads," he commented. But he also noted that, since the conference and partly because of it, new observations of the solar output, computer modeling efforts, and other approaches to problems of solar-terrestrial relations "are bringing us closer to the day when we can determine, in theory at least, whether these measured solar changes can have any significant effects in an atmosphere such as ours. In 1977, as in 1973 and 1895, we still don't know."

Common sense says that it is implausible to propose that the tiny tail of solar influences alone can wag the huge dog of the climate system. The sketchy and circumstantial evidence that has been put forward to support such proposals has generally failed to meet the rigorous standards of physical scientists like the Soviet meteorologist Andrei Monin, who wrote in 1972 of what he called heliogeophysics enthusiasts that:

> Most of the information concerning such an influence fortunately produces only an impression of successful experiments in autosuggestion; the hypotheses proposed concerning the physical mechanisms of the influence of solar activity on the weather lack convincing substantiation.

Monin used the word *fortunately* because he believed that proof of real solar-weather relationships would be almost a tragedy for meteorology, since it would demand that one be able to predict solar activity in order to predict terrestrial weather.

Nevertheless, many physical scientists are now agreeing that there is impressive evidence that certain weather and climate phenomena are linked with solar activity. If solar activity is indeed a significant factor in climatic fluctuation and change, scientific progress will not be made by trying to wish the problem away. Like other theories of climatic variation, this one deserves to be examined systematically and rigorously in a careful effort to test its validity.

Cars and trucks abandoned by their drivers during a blizzard that left highways virtually impassable line Interstate 65 near Lafayette, Indiana, at the height of the infamous Winter of '77. (United Press International.)

CHAPTER 5

Forecasts, Outlooks, Intelligent Guesses, and Conjectures

S OMETIME around the middle of January 1977, as record cold temperatures and heavy snow gripped the eastern and midwestern United States and drought grew steadily worse in the western states, the news media started referring to the "Winter of '77." The phrase had a sort of mythic ring to it, like an old man telling his grandchildren about things that happened long ago in the harsh and primitive days when he was a boy.

But the Winter of '77 was no myth. It was a bitter reality that inflicted great personal and economic hardships on many people who had thought they were living in a modern technological-industrial society that had shielded itself from most of the unpleasant impacts of weather and climate (Figure 1).

By the end of January, Chicago had hit a low temperature of −28 degrees Celsius (−19 degrees Fahrenheit) for its coldest day in this century. All-time low temperatures had been recorded in such diverse locations as Cincinnati (−32°C; −25°F) and Palm Beach (−3°C; +27°F). South Florida had snow for the first time in human memory, and an analysis of satellite photographs by the National Oceanic and Atmospheric Administration (NOAA) revealed that 65 percent of North America—about 6.4 million square miles—was snow-covered during January. This was the most extensive North American snow cover in the 11 years since NOAA began observing it with satellites. In mid-February, after searching official and unofficial weather records extending back into the eighteenth century, the National Weather Service announced that the Winter of '77 had been the coldest in the eastern two-thirds of the nation "since the founding of the Republic."

At least a dozen people froze to death in cars stranded by blizzards; altogether, more than 75 deaths were blamed on the cold weather. By early February, an estimated 1.5 to 2 million people had been thrown out of work by plant and business shutdowns caused by the natural gas shortage that resulted from the frigid temperatures. Public schools in many

FIGURE 1   Residents of Buffalo, New York, and other midwestern and East Coast cities that were buried by blizzards during the Winter of '77 had to dig their homes and cars out of great drifts built up by the combination of strong wind and heavy snowfall. (National Oceanic and Atmospheric Administration.)

areas were closed for periods ranging from days to weeks. Florida's fruit and vegetable crops were severely damaged by freezing temperatures, causing losses that one economist estimated could cost U.S. consumers as much as 7 billion dollars in higher food prices. The nation's heating bills were expected to run several billion dollars higher than in recent winters.

In the drought-stricken West, water rationing was instituted in communities in the San Francisco Bay area, and farmers in California's San Joaquin Valley were warned that their allotments of irrigation water might be cut by as much as 75 percent. In Colorado, where the winter snowpack in the Rockies provides a good part of the next summer's water supply for several western states, an emergency cloud-seeding program was mounted to try to augment snowfall on the half-bare mountain slopes. The attorney general of Idaho warned that his state would go to court if the State of Washington went ahead with a proposed cloud-seeding project over the Cascade Mountains, as he suspected that moisture that fell on Washington would be subtracted from Idaho's already scanty precipitation.

For weeks, the Winter of '77 was the biggest story in the newspapers and on television. Atmospheric scientists at the National Weather Service, the National Center for Atmospheric Research, and other organizations concerned with weather forecasting and research spent a great deal of time talking with journalists about what was happening to the weather.

It was not difficult for the meteorologists to explain *what* was happening in the atmosphere in the Winter of '77; *why* it was happening was not so clear. The cause of the cold in the East and the drought in the West was the behavior of the upper-level winds, as illustrated in Figure 2. The

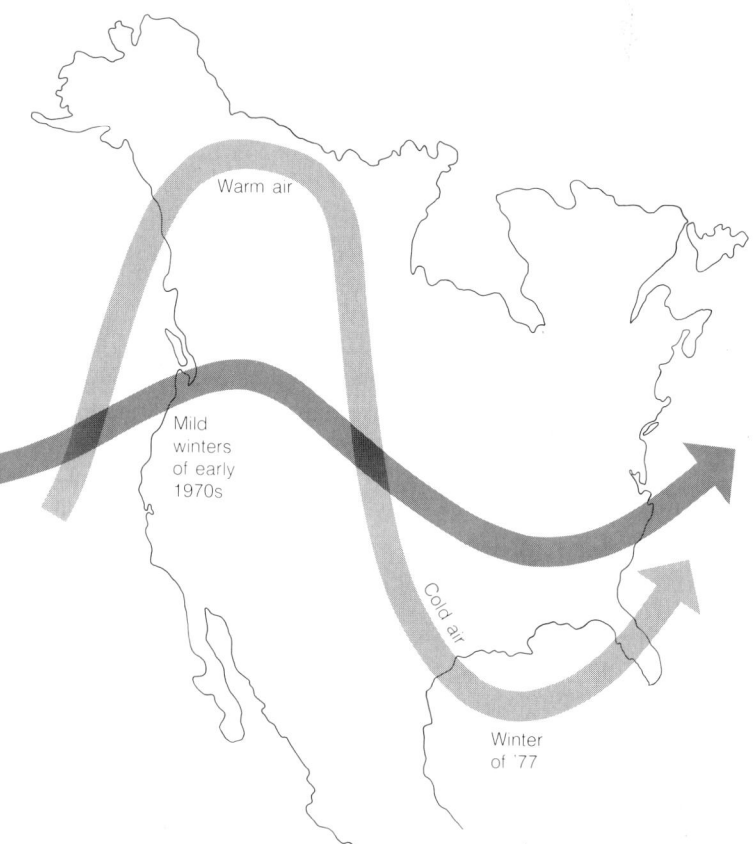

FIGURE 2 This simplified diagram of the upper-air winds over North America shows the pattern that resulted in the western drought and eastern cold and snow during the Winter of '77. (Henry Lansford.)

jet stream and other upper-level westerlies that steer mid-latitude weather systems had shaped themselves into a blocking pattern. A long wave swept northward as it neared the Pacific coast, producing a ridge of high pressure that carried warm air up toward Alaska. Then the upper-level winds swung sharply southward as they crossed the Rocky Mountains, bringing cold Arctic air far into the southeastern United States in a deep low-pressure trough. On some days in January, parts of Alaska had warmer temperatures than Miami; while the Florida citrus crop was freezing, an outdoor hockey match was canceled in Anchorage because the ice was melting.

The ridge-and-trough storm track over North America was not unusual in itself, but its persistence was. This blocking situation, hanging on for week after week, steered the winter storms north of the thirsty West and then flung them down across the eastern United States with ferocious intensity. Although it occurred in a completely different region, this pattern was similar to the one that caused Europe's drought in the summer of 1976, when a blocking high-pressure ridge hung over the Atlantic for many weeks.

Although atmospheric scientists could not explain definitively why the blocking situation persisted, some of them had theories about it. Jerome Namias of the Scripps Institution of Oceanography in California believes that ocean circulation and sea-surface temperatures play a key role in such large-scale weather patterns over North America. He said that an abnormally warm region in the Pacific Ocean off the coast of Washington and Oregon was probably related to the persistent upper-level ridge along the coast. Although he could not be sure of all the details, he felt that the sea-surface temperature anomaly and the ridge were produced by shifts in the warm tropical ocean current off South America called El Niño.

Two questions kept coming up when journalists interviewed the scientists: Why had the Winter of '77 taken so many people by surprise, and what could be done to predict future harsh winters and other disagreeable climatic developments?

According to the interviews, the Winter of '77 was not a complete surprise to many meteorologists. Donald Gilman, who heads the long-range prediction group at the National Weather Service, pointed out that his people had warned in Congressional testimony last fall that a severe winter was coming up. Jerome Namias, who preceded Gilman at the National Weather Service before joining the Scripps Institution in 1968, said that his group had also predicted a cold winter, but not as cold as it had in fact turned out. "I was a little too conservative," Namias said in late January. "Our forecast was for the coldest winter in perhaps 20 years, but now it looks as if 75-year-old records are going to be broken in many locations before the winter is over."

Stephen Schneider of the National Center for Atmospheric Research was not surprised by the severe winter. "You could say that I predicted a cold winter last summer," Schneider said, "But my forecast was strictly actuarial. After five mild winters we were due for a cold one and by chance we got it in 1977."

Even "Abe Weatherwise," who writes about weather for *The Old Farmer's Almanac*, predicted late in 1976 that the winter would be cold. So apparently a good many people whose business is predicting the weather were not greatly surprised by the Winter of '77. As might be expected, the long-range forecasters whose predictions had not come very close were not much inclined to discuss their lack of success, nor were they sought out by reporters who were covering the Winter of '77.

Hurd Willett, Professor Emeritus of meteorology at the Massachusetts Institute of Technology, has attracted a good deal of attention from the news media over the years with his long-range forecasting techniques based on apparent correlations between solar activity and terrestrial weather and climate. In a Sunday-supplement article published late in 1974, Willett was described as "an MIT professor with an uncanny record for correct long-range weather forecasts," and a number of his successful predictions over the last quarter century were presented as proof of his skill. The same article presented Willett's outlook for the winter of 1974–1975, which he predicted would be one of the coldest ever experienced east of the Continental Divide, with record-breaking low temperatures in Ohio and other midwestern states. In an interview published in *Technology Review* in January 1976, Willett made a similar prediction for the winter of 1975–1976. For the nation east of the Continental Divide, he said, "the current prolonged period of very warm weather should terminate before the end of November, to be followed by a prolonged spell of very cold weather, probably most severe during the mid-winter month of January, to give us a winter season markedly colder than normal." When both winters turned out mild, no reporters went back to interview Willett about his incorrect forecasts.

If either of these forecasts had been made for the 1976–1977 winter, Willett would have been in a good position to point triumphantly to their accuracy as proof of the validity of his forecasting techniques. However, wrong predictions are not very newsworthy, and meteorologists, like the rest of us, have a natural tendency not to talk nearly as much about the times they were wrong as about the times they were right. We are not singling Professor Willett out for special criticism; he has a distinguished record of notable contributions to atmospheric research. Our point is simply that we have not heard from all the meteorologists who did not anticipate that the Winter of '77 would be unusually cold, but only from those who did. And many of those, like Stephen Schneider with his "actuarial forecast," were simply saying that the odds appeared to favor a cold winter because the last five had been warm.

The lack of surprise expressed by many meteorologists at the Winter of '77 clearly came more from their readiness to expect almost anything from the weather than from any special foreknowledge of what was going to happen. This became obvious when reporters started asking about the next winter. Everyone backed away with great alacrity. Asked what the Winter of '77 could tell us about future weather and climate, Schneider answered honestly and succinctly: "In detail, nothing."

Other atmospheric scientists concurred. This year's cold winter in the eastern and midwestern United States did not mean that next winter would also be cold, they said, nor did it mean that next winter would be milder. It was almost like flipping a coin; if it comes up heads this time, that doesn't mean that it is either more or less likely to come up heads again next time. In short, it appeared to be impossible to find a reputable weather forecaster who was prepared to try to predict the winter of '78 a year in advance.

And yet there is no shortage of prophets, including some scientists as well as lay people, who are willing to predict the climate of the next several decades. Many of these prognoses have appeared recently in popular books about climate. The dust jacket of one, entitled *The Cooling* (Ponte, 1976), asks: "Has the next ice age already begun? Can we survive it?" *Hothouse Earth* (Wilcox, 1975), written by a physicist, includes a lengthy table showing which of the world's major cities will be submerged when the polar ice caps are melted by the global warming produced by continuing human use of fossil and nuclear fuels. And a third book, by a British science writer with an international reputation based on his books and BBC television scripts on scientific subjects, maintains that present climatic conditions are just about perfect for the sudden onset of an ice age that "could easily kill two thousand million people by starvation and delete a dozen countries from the map" (Calder, 1975). Many people find the gloomy assurance of these climatic Cassandras more depressing and confusing than the diffidence of the weather forecasters.

Not all atmospheric scientists are conservative in expressing their views on future weather and climate. We may read in today's newspaper than a prominent research meteorologist, speaking at a national meeting of a reputable scientific society, has warned that continued use of fossil fuels at the present rate of increase will almost certainly warm the earth significantly by the year 2000. But on tomorrow's television news, we are likely to see another scientist with equally impressive credentials explaining that the global cooling trend that began in the 1940s will probably continue for at least another few decades and may be a precursor of a new descent into glaciation.

Is it surprising that many people, including some decision makers who deal with national and international problems and policies on such climate-sensitive matters as grain reserves, energy resources, and foreign aid, have developed a tendency to stop listening when scientists start talking about future weather and climate?

But there are many potentially useful things that the atmospheric scientists *can* tell the decision makers. One big communication problem lies in the fact that, when meteorologists or climatologists start talking about the future, it is sometimes difficult to tell whether they are presenting a forecast, an outlook, an intelligent guess, or a conjecture that they find provocative even though it is not supported by any very convincing evidence.

What is the current state of weather and climate prediction? Can atmospheric scientists give decision makers forecasts that are both skillful enough and specific enough to serve as useful inputs to social, economic, and political decision-making processes? And can outlooks, intelligent guesses, and even conjectures about future weather and climate be useful in decision making if they are clearly labeled as what they are and are not taken as forecasts?

Let us consider just what *is* and *is not* possible in weather and climate forecasting, using existing skills and techniques, and then look at potential improvements that may be found through research in the next decade or so. As we did in our discussion of weather and climate mechanisms, we shall begin with short-term weather phenomena that are reasonably well understood and proceed from there to the climatic fluctuations and changes that we do not understand very well.

Although scientific weather forecasting really dates back about a century, attempts to predict the weather accurately have been made for many centuries. One of the earliest efforts to develop a systematic approach to weather forecasting was undertaken by Aristotle's pupil Theophrastus, who in 300 BC wrote the *Book of Signs*, which would be the definitive work on weather forecasting for the next 2000 years. It was a collection of more than 200 natural signs that Theophrastus had concluded were significant predictors of weather. Many of these signs are still part of our weather folklore. The old mariners' jingle "Red sky at morning, sailor take warning/ Red sky at night, sailor's delight" may be found, in more prosaic language, in the *Book of Signs*.

Over the centuries, the approach used by Theophrastus remained the basis for attempts to forecast weather. A variety of apparently unrelated natural signs were used to predict various weather phenomena, but there was no successful attempt to relate these phenomena to one another as different aspects of a great natural system.

In the seventeenth century, as new instruments appeared for measuring the properties of the atmosphere, weather observers began to build regular records of temperature, humidity, barometric pressure, as well as wind and precipitation, laying a foundation for scientific weather prediction. By the middle of the eighteenth century, scientists had begun to understand that the weather in a particular location did not occur in isolation but was linked to weather conditions in other places. In 1743, Benjamin Franklin discovered an important principle of atmospheric behavior quite by accident. He observed a violent coastal storm in Philadelphia on October 21, with winds from the northeast. He assumed that the storm had come from that direction but was surprised to learn that the next day Boston experienced similar severe weather. He concluded that the winds within a storm do not necessarily blow in the direction in which the storm is traveling. This was an important insight into the nature of cyclonic storms, which rotate in a counterclockwise direction around a low-pressure center as the whole storm moves slowly across the face of the earth, usually from west to east.

The knowledge that large-scale weather systems move across the land did not have much practical value for forecasting as long as weather information could travel no faster than the weather itself. The invention of the telegraph marked the beginning of synoptic meteorology—the synthesis of many observations made simultaneously at widespread locations to depict the state of the atmosphere over a large area. By 1850, weather reports were being telegraphed from dozens of stations in the United States to the Smithsonian Institution in Washington, DC, where weather maps were published for the area covered by the network.

The Cincinnati Observatory made the first real weather forecasts in the United States; they were known as probabilities. The U.S. Army Signal Service issued the first official U.S. government forecasts and established a weather service designed to provide storm warnings for shipping on the Great Lakes. During the last part of the nineteenth century, many other nations were also establishing national weather services. Weather forecasting in the United States and Europe in these early years consisted mainly of watching for stormy weather and assuming that it would move eastward. There was little understanding of the mechanisms of the birth, growth, and decay of the weather systems themselves.

In the first years of the twentieth century, remarkable advances in weather theory were made by a group headed by Vilhelm Bjerknes at the Norwegian Geophysical Institute. Their most important contribution to modern meteorology was probably the development of the polar-front theory, which accounts for the birth of cyclonic-scale storms by collisions between warm and cold air masses. Bjerknes also proposed an approach to the physical behavior of the atmosphere that is fundamental to much of the theory and practice of meteorology today. In 1904, he suggested that weather forecasting could be considered "a problem in mechanics and physics." Stated simply, Bjerknes' proposition was that the atmosphere is a fluid, subject to the laws of physics. If its physical behavior can be expressed mathematically, it should be possible to state the present condition of the atmosphere numerically and to predict future weather by simulating its behavior with a series of mathematical computations.

Bjerknes was not successful in his attempts to develop numerical techniques for solutions to the problem of the evolution of weather. An English scientist, Lewis F. Richardson, came closer than Bjerknes, but he was defeated by the incredible amount of data and number of computations needed. To put into operation a numerical forecasting system that he proposed in 1922, Richardson would have needed weather data from 2000 stations distributed uniformly over the earth. And to perform the computations involved in his numerical prediction system, he would have needed 64,000 calculating machines operated 24 hours a day by 64,000 human operators.

Almost 25 years later, after the advent of the electronic computer, Richardson's theories were tested and found to be generally sound. In 1946, John von Neumann and his colleagues at Princeton University used a computer called MANIAC (Mathematical Analyzer, Numerical Integrator, and Computer) to develop Richardson's ideas into numerical models.

Von Neumann's work led to the development of numerical models of the atmosphere similar to those used today by the National Weather Service in making regular operational forecasts. Efforts to develop more realistic models of the global atmospheric circulation that may eventually be used to extend the time-scale of accurate weather forecasts from the present two or three days to a week or more are going on at the NOAA Geophysical Fluid Dynamics Laboratory in Princeton, the National Center for Atmospheric Research, the University of California at Los Angeles, and other research centers in the United States and abroad.

What is a weather forecast, in terms of present-day knowledge of the mechanisms of the atmosphere that produce our weather? When meteorologists uses the term *forecast,* they usually mean a fairly specific and detailed prediction of the weather for the next day or two. For example, if you live within 40 miles or so of Denver and have a radio that can be tuned to the frequency of 162.55 megahertz, you might hear something like this on a Sunday evening in February:

> These are the current forecasts. First, for the Denver metropolitan area, a chance for snow showers during the remainder of tonight with accumulations generally one inch or less over the eastern portions of the metropolitan area and one to three inches near the foothills. Clear to partly cloudy Monday through Tuesday, cooler on Monday. Lows tonight and Monday night in the mid-20s, highs Monday and Tuesday 50 to 55. Winds now 10 to 25 miles an hour and gusting, but decreasing after midnight. Probability of measurable precipitation is 30 percent tonight.
>
> The outlook for Denver Wednesday through Friday, continued dry and mild, highs in the 50s and lows in the 20s.

Note the shift in specificity in this broadcast from the National Weather Service in Aurora, Colorado—a Denver suburb—from the *forecast* for the next two days, which deals in temperatures to within about 5 degrees, precipitation in inches, and wind velocities in miles per hour, to the much more general *outlook* for the remainder of the upcoming five-day period.

This particular forecast did not come out too badly. In Boulder, which lies right at the beginning of the foothills, we had just over 3 inches of snow that Sunday night. The low temperature Monday night was 18, and the high Tuesday was 47, both only a few degrees colder than predicted.

The skill of the weather forecaster drops off badly as the range of the forecast grows longer. Although the nomenclature of weather forecasting is not as precise as it might be, most meteorologists distinguish between *short-range* forecasts of up to 48 hours, *extended-range* forecasts of up to about 5 days, and *long-range* forecasts of up to a month or even longer. The last two, which tend to be general in nature, are often called *outlooks*.

How are these forecasts and outlooks made? It should be clear even to those whose knowledge of weather forecasting is limited to brief weather reports on television news programs that present-day weather forecasting involves considerably more science than intuition. Even television weather reporters toss around references to cold fronts and jet streams and illustrate their comments with weather maps and satellite photographs (Figure 3).

As recently as the mid-1950s, the intuition and subjective skill of the individual weather forecaster were the most important ingredients of an accurate forecast. In those days, forecasters began with a synoptic

FIGURE 3 Meteorologist Bob "Sunny" Roseman analyzes weather patterns for viewers of KBTV Television in Denver. Most television weather forecasts, although presented in terms that the lay person can understand, are solidly based on advanced technology that uses satellites, computers, and other sophisticated tools to analyze and predict the behavior of the atmosphere. (KBTV Television, Denver, Colorado.)

weather map showing surface-level pressure patterns, temperatures, and winds, along with charts that depicted conditions at several upper levels of the atmosphere. They would then proceed to extrapolate tomorrow's weather from today's by starting with the past speed and direction of existing cyclonic systems and drawing on their own empirical knowledge of how those systems were likely to be affected by the large-scale conditions that prevailed around them. The greater the skill and experience of the forecaster, the higher the probability that tomorrow's real weather would match the synoptic forecast reasonably well.

By the late 1950s, research meteorologists had built on the work of von Neumann and his colleagues at Princeton to develop numerical weather prediction techniques that could be used to forecast large-scale pressure, temperature, and winds by computer with reasonable accuracy.

These early numerical forecasts still required the forecaster to use skill and knowledge of local influences to predict smaller scale features and to add clouds and precipitation to the forecast. Even today, when a computer at the National Meteorological Center at Suitland, Maryland, produces a large-scale numerical forecast that is transmitted to all National

Weather Service forecast offices across the country, the forecast that reaches the public includes a large contribution from skilled human forecasters who fill in the broad outlines of the numerical forecast.

Because regional and local weather are so strongly influenced by the global-scale behavior of the atmosphere, most of the nations of the world exchange weather data through the World Meteorological Organization (WMO), a specialized agency of the United Nations. The history of WMO goes back to 1873, when its predecessor organization, the International Meteorological Organization, was created by representatives of 20 countries who met in Vienna and agreed that the national weather services of the member nations would cooperate to exchange weather data and engage in other international meteorological activities.

Today, 145 member nations operate the World Weather Watch, a global meteorological system composed of the facilities and operations of the individual national weather services. The three principal components of this system are a global observing system, a global data-processing system, and a global telecommunications system.

The basic weather observations in this system are made four times each day, when simultaneous measurements are calculated with standardized instruments at more than 9000 weather stations around the world and by several thousand ships at sea. Upper-air measurements are also made with rawinsondes—weather balloons that transmit data back to the ground by radio—released at more than 500 stations. All these synoptic observations go first to national and regional data centers, then to three World Meteorological Centers, located at Washington, Moscow, and Melbourne. Radar, aircraft, and satellite observations are also part of the global observing system. From the world centers, data sets go back down to the regional and national centers to be used in preparing synoptic charts that give the forecaster a three-dimensional picture of the current (or very recent) state of the atmosphere.

The data also go into computers to provide the initial conditions for numerical forecasts. By solving the equations of the atmospheric model over and over again, the computer makes the weather "happen" in its electronic circuits faster than it happens in the real atmosphere, thus producing a numerical forecast.

From weather centrals such as the National Meteorological Center in Maryland, the charts, data tabulations, and numerical forecasts are transmitted over telephone lines to facsimile machines that reproduce them at National Weather Service regional and local forecast centers. There, the state and local forecasts are prepared and disseminated to the public. The Denver metropolitan area forecast that told us to expect 1 to 3 inches of snow near the foothills on that Sunday night in February was one end-product of this massive global task of weather observation, data transmission and collection, plotting and analysis, and, finally, preparation and dissemination of the weather forecast for the next 24 hours and the outlook for the next 5 days.

How accurate is that forecast likely to be for Denver, or for New York, London, Berlin, Karachi, Dakar, Bangkok, and all the other places where people want to know what the weather will be like tomorrow, next week, or a month from now? The answer depends primarily on two factors: the time-scale, or range, of the forecast and the geographical location for which it is being made.

Not long ago, the American Meteorological Society adopted a policy statement on weather forecasting designed to "describe the present weather-forecasting capability of the meteorological profession." This statement does not have any flavor of professional optimism or self-serving enthusiasm; it seems to be a hard-headed attempt to assess what is and is not possible in weather forecasting. Here is its summary of average levels of skill for forecasts of various periods in mid-latitude regions of the Northern Hemisphere:

1. *For periods up to 48 hours:* weather forecasts of considerable skill and utility are attained. Detailed forecasts of weather and its changes can be made for the first 36 hours. . . . In periods up to 48 hours, skill is at a maximum in predicting the motion and general effects of weather systems having dimensions of 1000 kilometers [620 miles] or more. However, small-scale features imbedded in these systems cause hour-to-hour variations in weather that are difficult to predict, especially for local areas with irregular topography. . . .

2. *For periods up to 5 days:* daily temperature forecasts of moderate skill and usefulness are possible for periods extending to about 5 days. Precipitation forecasts to 3 days, at an equivalent level of skill, can be made, but the skill drops to marginal levels on the fourth and fifth days.

3. *For periods from 5 days to 1 month:* average temperature conditions can be predicted with some slight skill. Day-to-day or week-to-week forecasts within this time range have not demonstrated skill. There is some skill in prediction of total precipitation amounts for periods of 5–7 days in advance; skill for longer periods is marginal.

4. *For periods of more than 1 month:* skill in day-to-day forecasts is nonexistent and skill in seasonal outlooks and climate forecasts is minimal at the present time.

For locations that are not in the northern temperate zone, forecasting skill is much lower. A serious problem in the Southern Hemisphere is lack of data. With its huge expanses of ocean and many developing countries that do not have advanced national weather services, the Southern Hemisphere lacks the network of weather stations needed to provide the synoptic description of the present state of the atmosphere that is the starting point for forecasting its future states. In the tropics, the data problem is compounded by a lack of detailed understanding of the mechanisms and behavior of tropical weather systems. The lack of Southern Hemisphere data and understanding of tropical meteorology also affects forecasting skill in the mid-latitudes of the Northern Hemisphere for periods longer than 48 hours or so, as the global behavior of the atmosphere becomes the dominant influence in extended and long-range forecasting.

These two interrelated problems in weather forecasting—lack of data and lack of understanding—are being attacked through a double-pronged international effort. We have already mentioned the World Weather Watch and its efforts to improve global weather observation and forecasting. The Global Atmospheric Research Program (GARP) is a joint undertaking of the World Meteorological Organization and the International Council of Scientific Unions (ICSU). Conceived in the 1960s and launched in the 1970s, GARP is designed to develop an understanding of the global behavior of the atmosphere that will extend the time-scale of accurate, large-scale forecasts to its theoretical limit, which is probably about a week but could conceivably be as much as two weeks.

The first large-scale field project in this international undertaking was the GARP Atlantic Tropical Experiment (GATE), conducted during a 100-day period in the summer of 1974. GATE, which involved some 4000 people from 70 nations, was probably the largest and most complex scientific field experiment ever undertaken. The researchers used 39 ships, a dozen aircraft, nine satellites, and a variety of buoys, radars, balloons, and land-based weather stations to gather data from a huge expanse of tropical atmosphere, ocean, and land centered on an area of the tropical Atlantic west of the GATE field headquarters at Dakar, Senegal.

GATE was a remarkable exercise in international scientific cooperation. The director of the experiment was a U.S. scientist, Joachim Kuettner, and the deputy director was Yuri Tarbeev of the Soviet Union. The ships came from Brazil, Canada, France, East and West Germany, Mexico, the Netherlands, the Soviet Union, and the United States. Eight of the research aircraft came from the United States, two from the Soviet Union, and one each from France and the United Kingdom.

A primary goal of GATE was to define the role that weather systems known as tropical cloud clusters play in the large-scale circulation of the atmosphere. The end-product of the experiment was a mountain of data—literally tons of it—recorded on magnetic tape, photographic film, and other media. The data have been archived at centers in England, France, Germany, the Soviet Union, and the United States, where they are accessible to scientists from these and other nations. Although new increments of knowledge about tropical weather systems and their role in the global behavior of the atmosphere began to emerge even while the field work was still going on, five years or more were expected to pass before the analysis and interpretation of the GATE data were complete.

In the meantime, another field effort, the First GARP Global Experiment (FGGE), was scheduled to get underway in the late 1970s. Unlike GATE, which put an armada of ships, aircraft, and other observing platforms into the field for a short period of intensive observation of a comparatively small area, FGGE was designed to observe the atmosphere over the entire globe intensively for a year, using satellites, conventional weather stations, ocean buoys, and other observing tools and platforms.

These GARP field research programs are complemented by theoretical research aimed at improving numerical models used in research and prediction. As the field observations are used to make the models more realistic, the models can identify both critical gaps in data and particularly sensitive parts of the weather and climate system. By sometime in the 1980s, the GARP effort is expected to provide the theoretical basis for extending large-scale weather forecasting to its outer limit in time and to provide greatly improved knowledge of interactions within the climate system.

Numerical forecasts of specific details of weather—temperatures, pressure, winds, and other elements on a particular day—almost certainly will reach a limit somewhere between five days and two weeks, no matter how good the data sets and models are. However, forecasts of climate rather than weather—of average conditions over a given area and period of time rather than specific values—can probably be made for considerably longer periods. The present National Weather Service outlooks for more than five days are climate forecasts in this sense. The 30-day and 90-day outlooks are published widely by newspapers and magazines, which often call them forecasts even though they are very general (see Figure 4). How accurate are these long-range outlooks?

The 30-day and 90-day outlooks are simply maps of the United States overprinted with different patterns to indicate where the temperature and precipitation are expected to be below normal, near normal, above normal, or too close to call (Figure 5). But these designations tend to be rather vague for most practical purposes. For example, the 90-day tem-

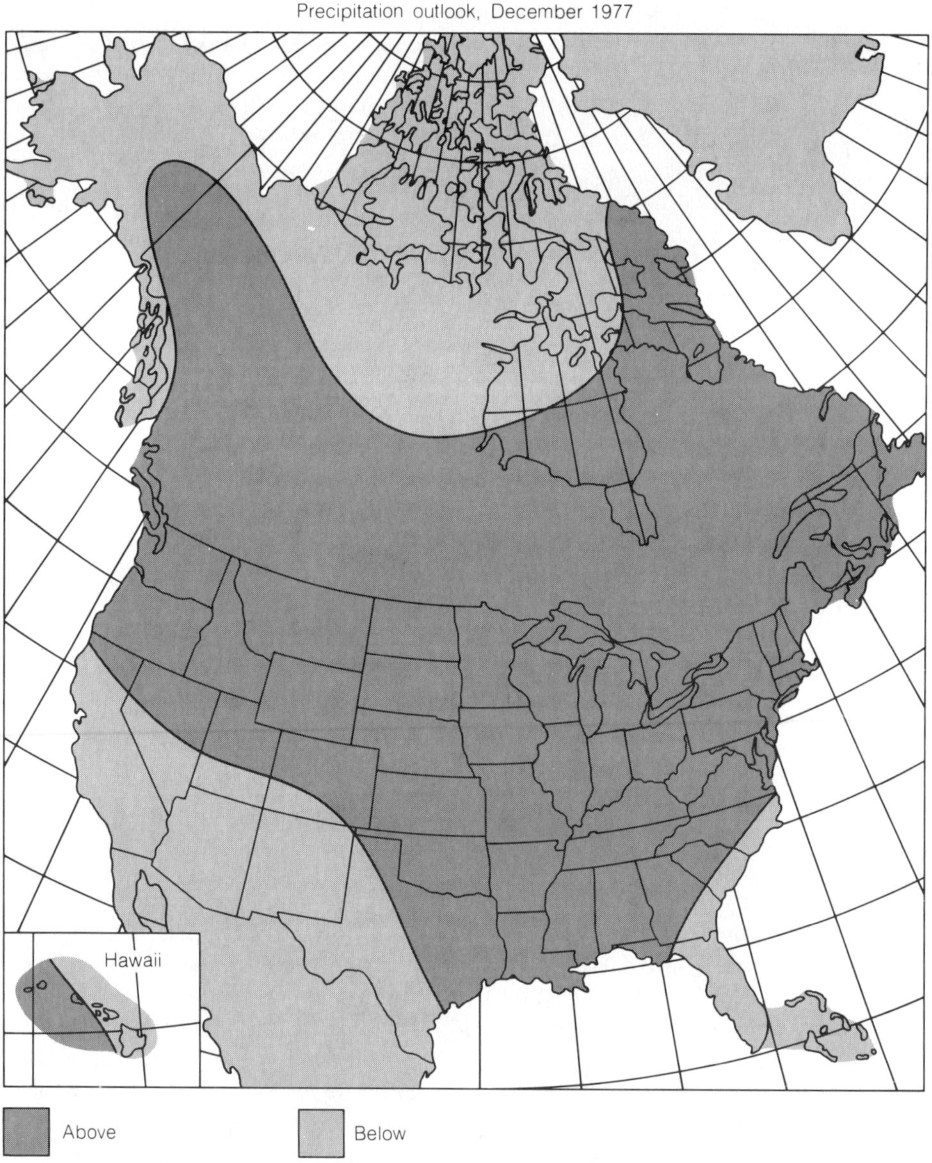

Precipitation outlook, December 1977

Hawaii

☐ Above   ☐ Below

FIGURE 4   The 30-day outlooks issued by the National Weather Service show
where both temperature and precipitation are expected to be above, below,
or near normal. (National Weather Service.)

Temperature outlook, December 1977

Hawaii

Above   Below   Near normal

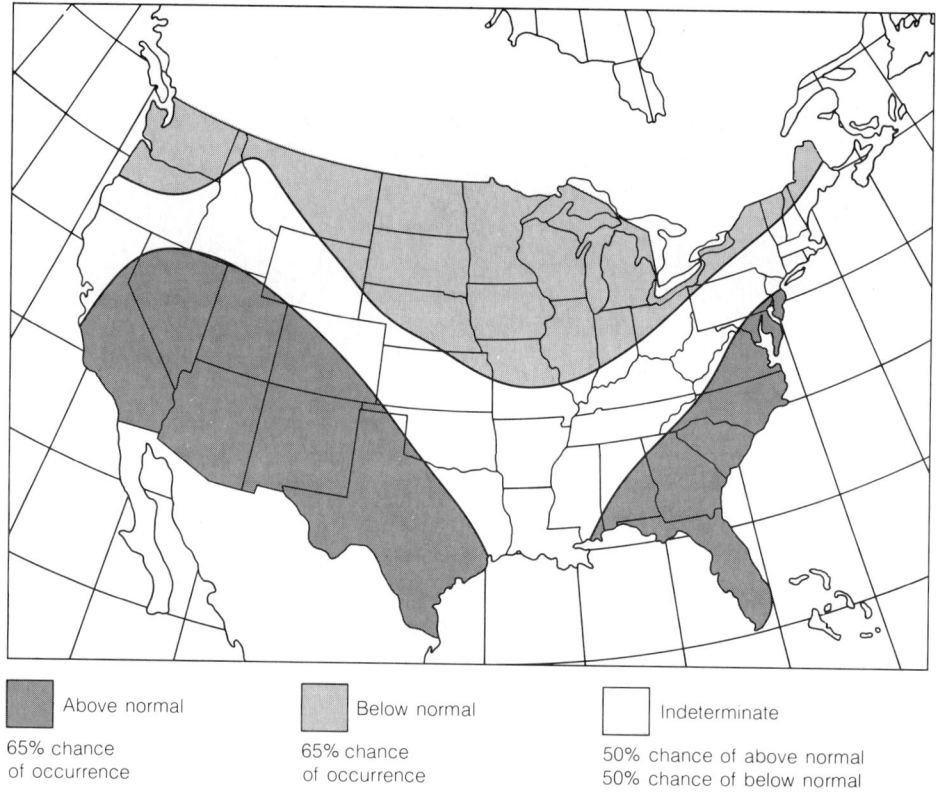

Above normal	Below normal	Indeterminate
65% chance of occurrence	65% chance of occurrence	50% chance of above normal 50% chance of below normal

perature outlook that appeared in *U.S. News and World Report* for December 13, 1976, indicated that south of a line running from Rhode Island roughly southwestward toward New Mexico, average temperatures from December through February would be "cooler than normal." Many people who lived through the Winter of '77 in that region would agree that this was the understatement of the century.

Long-range outlooks, for a month or more, are made by techniques that are much less precise than the mix of numerical and synoptic forecasting that is used for predicting weather for the next few days. At one extreme of credibility in long-range forecasting are almanac publishers and others who claim to predict the weather a year or more in advance by using folk wisdom of various kinds. The general nature of most of these "forecasts," combined with a little luck and intuitive skill, results in apparent success frequently enough to maintain a loyal following for Abe Weatherwise and his kind. Some private weather forecasters,

FIGURE 5 The National Weather Service 90-day outlooks are for temperature only, and they are considered to have only 65 percent probability of being correct. (National Weather Service.)

usually claiming that they possess secret techniques unknown to the National Weather Service and other forecasters, profess to be able to predict weather months in advance, but the evidence for their claims has not been convincing.

A few reputable long-range forecasters, such as Jerome Namias and Donald Gilman and his colleagues at the National Weather Service, do not make extravagant claims, but they admit that they are pushing the methods that they use beyond the limits of consistently demonstrable success. Namias, who founded the Extended Forecast Division of the National Weather Service and headed it for many years, developed most of the basic techniques that are used in putting together the 30-day and 90-day outlooks. He admits that the accuracy of the 90-day temperature outlooks is "nothing to brag about."

"The success of these seasonal outlooks is better than simple climatological probability," Namias says, "but if you count too much on the accuracy of an individual 90-day outlook, you're liable to get burned."

Some atmospheric scientists believe that numerical modeling offers the only chance for improving the success of extended and long-range forecasts. Namias admits that numerical prediction may offer the ultimate long-term solution, but he feels that it would be a mistake to neglect other approaches to the problem. He believes that useful results may be found much sooner by continuing to study statistical and empirical relationships among different features of the climate system. Namias points out that the national weather services of England, Japan, and the Soviet Union, as well as that of the United States, have demonstrated modest but promising progress in long-range prediction by such techniques.

In the statistical approach, according to Namias, it is taken for granted that the average prevailing wind and weather patterns for one month, together with snow cover and other associated abnormalities of sea temperatures and land surfaces, largely determine how the general weather patterns are going to develop during the following month.

Along with a number of other prominent climatologists, such as Hubert Lamb of England and Hermann Flohn of West Germany, Namias believes that new techniques for predicting some climatic fluctuations may come from studying "teleconnections"—correlations between simultaneous climatic anomalies such as the Soviet drought, the Indian monsoon failure, and the peak of the Sahelian drought, which all occurred in 1972. Namias says that there are strong indications that such anomalies are not distributed randomly but have interrelationships that can be discerned by rigorous analysis of existing data.

Namias points out that the atmosphere has a short "memory"—its features and interactions change rapidly, so that today's atmospheric conditions may have very little direct connection with the weather three or four weeks from now. But the oceans, which are coupled with the atmosphere through many exchanges of moisture and energy, change on a much longer time scale. Namias believes that "the ocean is probably one major source of memory imposed on the atmosphere, particularly in the tropics." In a paper published in 1972, he described this ocean-atmosphere relationship like this:

> In the past decade, research has shown that the thermal state of the oceans, especially the temperatures in the upper few hundred meters, varies considerably from month to month and year to year, and that these variations are both cause and result of disturbed weather conditions over areas thousands of miles square. Thus the prevailing wind systems of the globe—the westerlies, the trade winds, and the jet streams—may be forced into highly abnormal patterns with concomitant abnormalities of weather. Because these reservoirs of anomalous heat in the ocean are deep, often up to 500 meters, and may last for long periods of time, the atmosphere can be forced into long spells of "unusual" weather, sometimes resulting in regional droughts or heavy rains over periods ranging from months to seasons, and even years or decades.

This is the kind of ocean-atmosphere interaction that Namias believes was responsible for the Winter of '77 over North America. He maintains that empirical studies of teleconnections may reveal recurring relationships between certain phenomena in the oceans and atmosphere that can be used as predictors of climatic fluctuations and anomalies, even though the mechanisms that are involved are not sufficiently well understood to be incorporated into numerical models. Namias believes that, by use of statistical and empirical techniques in conjunction with models, significant improvements in long-range climate prediction on scales of six months to a year can be achieved, even though the models themselves are not yet sufficiently sophisticated to make such predictions by numerical modeling techniques.

"I'm optimistic," Namias says, "but still aware that there may be no way to make a reliable prediction beyond a season in advance. The atmosphere itself may not know what it's going to do next—from one initial state there may be half a dozen different ways it could go."

There is not much doubt that accurate climate forecasts issued a year, six months, or even a season in advance could provide great benefits for some users. But the question of just what dividends could be expected from climate forecasts of particular time ranges and specificities has not been examined very carefully. There has been a somewhat uncritical tendency to assume that if we could foretell the climatic future, we could respond to it rationally and appropriately. But in terms of society's ability to respond to weather and climate, forewarned is not necessarily forearmed.

Here is a simple example. Many wheat farmers on the U.S. high plains do not harvest their own crops but use the services of "custom cutters"— contractors who follow the wheat harvest north from the Texas Panhandle to the plains of Canada with fleets of combines and trucks and crews of operators. If a high-plains farmer has a couple of thousand acres of ripe wheat waiting to be harvested, but the custom cutter won't arrive until the day after tomorrow, a forecast for severe hail tomorrow won't be of any practical use, regardless of its accuracy. The constraint of the custom cutter's schedule has closed off the farmer's options; the only thing possible to try to save the wheat from being battered to chaff and stubble by hailstones is pray that the forecaster was wrong this time.

On a larger scale, a variety of societal constraints might limit the options for responses to climate forecasts. Michael Glantz, a political scientist, recently completed a study (1976a) of "the social, political, economic, and environmental implications of a credible and reliable long-range climate forecast." This study, which was sponsored by the International Federation of Institutes for Advanced Study (IFIAS), began with the assumption that precipitation and temperature could be predicted a year in advance in terms of whether they would be higher or lower than long-term averages.

In one of his case studies, Glantz assumed that such a forecast was available for the Sahel in October 1972, predicting the severe drought of 1973 that came to the Sahel as a devastating climax to the series of dry years that began in 1968 (Figure 6). During 1975 and 1976, Glantz asked officials of Sahelian governments, scientists who were familiar with problems of the region, and a variety of other experts what they would have done or recommended if an accurate climate forecast for the Sahel had been available in the fall of 1972, six months before the start of the 1973 growing season. The experts were asked to assume that there would have been no constraints on carrying out their recommendations. Once he had compiled these suggestions, Glantz considered the political, economic, social, and environmental constraints that might act to prevent carrying out the recommendations of the experts about how to apply the forecast effectively (which he called "what ought to be" as distinguished from "what is"—the real situation dictated by the constraints).

One of the most frequent suggestions was that the Sahelian govern-

ments assess the carrying capacity of the rangelands so that the size of the herds of cattle could be reduced to prevent overgrazing during the anticipated period of low precipitation. Herders would then be required or requested by the government to cull their herds, keeping the best cattle in order to improve the quality of the herds when their size was permitted to increase again. The herders could use the money from the sale of the cattle to buy grain for use during the drought. The meat could be consumed locally, dried or frozen for future use, or exported to markets in West African coastal countries where meat is scarce.

Other "what ought to be" suggestions involved encouraging temporary migration of some of the pastoral people to other areas to find cash employment, establishing planting schedules that would be adjusted to the drought conditions, producing fodder to supplement the sparse forage, setting price controls—a maximum on grain and a minimum on livestock—estimating impending food and medical needs, and arranging to purchase food stocks and seek foreign aid before a regional emergency existed.

But many of these apparently sound suggestions appeared destined to founder on the rocky shoals of "what is" in the Sahel. Assessing the carrying capacity of the rangeland under various climatic conditions requires detailed ecological data and knowledge that are not available. The bureaucracies of the Sahelian governments are fragmented into ministries and departments that operate in an atmosphere of rivalry, poor cooperation, and frequent corruption that is not conducive to efficient and effective administration of government policies. The traditional desire of the herders to own as many cattle as possible would be a serious obstacle to a rational stocking plan based on a realistic assessment of the carrying capacity of the land under drought conditions. Transport systems in the Sahel are inefficient and badly suited to transporting either live animals or fresh meat to market. Grain storage facilities are inadequate, and stored grain is subject to heavy losses to spoilage and pests.

Glantz tentatively concluded that, "given the present social, economic, and political settings in which a potential technological capability would be used, the value of that capability (in this case a long-range climate forecast, even a perfect one) would be disappointing." In other words, even if the people of the Sahel had known in the fall of 1972 that the drought would continue in 1973, the societal and environmental impacts probably would have been as devastating as they were in the absence of such a forecast (Glantz, 1976a).

Although conventional wisdom tells us that a technological-industrial society with a comparatively strong and efficient central government could respond much more effectively to climate forecasts than the developing nations of the Balkanized Sahel, there are constraints to be found in developed nations as well. Suppose, for example, that in the fall or winter farmers in the northern part of the U.S. corn belt are warned that there is a high probability of early frost next autumn. Theoretically, it should be possible for the farmers to plant faster maturing varieties of corn instead of the higher yielding varieties that require a longer growing season. But the seed corn that they plant was grown the previous year. If the seed growers did not anticipate a high demand for corn suited to a shorter growing season, the seed may not be available, and the farmer will have to plant the usual variety and hope for the best. Unless the forecast provides an extra year's lead time for the seed growers, it could be as useless as the hail forecast for the wheat farmer who has no way to harvest the wheat.

On the other hand, if the seed growers and farmers are warned that early frosts will become more probable in future years—what Schneider calls an actuarial forecast—then there are several workable options. If an accurate seasonal forecast is possible, the seed growers could stock large supplies of more than one variety of corn, allowing the farmers to make a last-minute decision. Or the farmers could decide to plant lower yielding but faster maturing varieties year in and year out, accepting lower yields in what might have been bumper years on the principle that they will be balanced out in the long run by reasonably good yields even in the bad years.

The point is that, no matter how good a weather or climate forecast may seem to the meteorologist in terms of its skill or accuracy, its value to the user depends on its relevance to the choices that are available. And in the absence of forecasts or outlooks precise enough to satisfy a meteorologist, the farmer is likely to make some hard and important decisions on the basis of intelligent guesses or conjectures about future weather and climate that fall far short of the atmospheric scientist's rigorous standards of acceptability.

For example, every dryland farmer in the high plains probably knows by now that severe drought has struck the region about every 20 to 22 years for the last 160 years. Some scientists believe that there is an inherent periodicity to these droughts, whether or not it can be related to the sunspot cycle. Thus the belief that another high-plains drought would occur in the mid-1970s was not a forecast or an outlook, but it was certainly an intelligent guess. It differed substantially from the conjecture of the early homesteaders that "rain follows the plow," as it was supported by a long period of record rather than just a few years' observations. And even though the meteorologist might protest that, in the absence of a physical explanation, it was not scientifically sound to assume that the drought of the 1970s would occur, acceptance or rejection of this assumption was about the only basis on which the farmer could decide whether or not to plow under a shelterbelt of Russian olive trees or put some land into pasture instead of dryland wheat.

Even conjectures can be of some use in making climate-related decisions in the real world, provided they are not completely wild. Many scientists feel that Reid Bryson's contention that the global climate will probably continue to cool for some time is definitely a conjecture rather than a forecast, an outlook, or even an intelligent guess. But some of them also feel that Bryson has done society a useful service by making the public aware that such a climatic development is at least within the realm of possibility. Murray Mitchell, who has contested Bryson's views publicly and with some acerbity over the last few years, shares his concern about the impact of climate on food supplies. Mitchell does not believe that anyone can predict the future course of the climate, but he feels that climatic variability is almost certainly going to affect agricultural production.

"We've been extraordinarily lucky, because the weather in the major crop-growing areas of the United States has been uniformly good for so many years—so uniformly good that it's almost a fluke, in a climate that over the long run has varied much more than that," Mitchell says. "To make plans as if the climate were going to continue so favorable for U.S. agriculture would be absolutely irresponsible. From the agricultural productivity point of view, the climate's not going to get any better. It can only get worse."

Whether you consider that an intelligent guess or a conjecture, Mitchell is clearly saying that a lack of precise knowledge about what the climate is going to do is not a valid reason to ignore the possibility of future climatic anomalies for which we may be as unprepared as we were for the Winter of '77. A forecast would be the best basis for making climate-related decisions on matters such as grain reserves, energy resources, and foreign aid. An outlook would be second best. But until we have forecasts and outlooks for climatic variations that we can expect a month, a year, or a decade from now, we had better make the best use that we can of intelligent guesses and even conjectures about future weather and climate.

CHAPTER 6

Climate, Crops, and Consumers

I N CHAPTER 4 we discussed the climate system, made up of the interacting internal elements of atmosphere, oceans, and ice masses, and
influenced by external boundary conditions such as the input of
energy from the sun. Now we shall consider another system of interacting elements, one that includes social, political, and economic as well as
physical components. We might call it the climate-crop-consumer
machine.

The Winter of '77 provided a simple but dramatic example of how the
climate-crop-consumer machine operates on a national scale. Two of the
major winter vegetable–producing regions in the United States, Florida
and California, were hit respectively by frost and drought, with serious
damage to the vegetable crops. During February, fresh vegetable prices
jumped a record 20.9 percent (Figure 1). This increase, along with some
other factors, resulted in an overall February increase in retail food prices
of 2 percent, and the consumer price index rose 1 percent in the largest
inflationary surge in almost three years.

A more complex and large-scale response of the system to a climatic
perturbation occurred in 1972, when weather all around the world
behaved in ways that were highly unusual by comparison with the
"normal" climatic conditions of the previous 15 years or so. In Chapter 2
we reviewed some of the climatic anomalies of 1972—the harsh winter
followed by a dry, hot summer in the Soviet Union; monsoon failure in

The 1975 wheat harvest was the first in a series of
bumper crops of grain in North America that resulted
in growing grain surpluses and dropping prices.
(United States Department of Agriculture.)

FIGURE 1   Fresh vegetable prices in the supermarkets rise very quickly when bad growing weather hits Florida or southern California. (Henry Lansford.)

India; floods in the U.S. Midwest; and widespread drought in West Africa, South America, Australia, and some other regions.

This generally unfavorable weather resulted in the first drop in global food production since the years just after World War II. Total world food production actually declined by less than 2 percent in 1972, but the drop followed a steady increase, totaling more than 25 percent, that occurred between the mid-1960s and 1971, largely as a result of the technological innovations of the Green Revolution. As a 1974 U.S. Department of Agriculture (USDA) report put it: "The crop shortfalls were particularly unsettling because they broke the growing confidence about overcoming the world food problem that had emerged in the period after 1966."

Major crop failures in the Soviet Union and India were accompanied by mediocre harvests in Canada, Argentina, and Australia. But the U.S. grain harvest was good—a near-record 224 million tons—and the United States already had a reserve of more than 70 million tons of wheat and feed grain on hand. The North American "breadbasket" was prepared to export enough grain to feed the hungry people of other countries, as it had many times in the last quarter century.

Among the grain buyers from other nations who came to the United States was a team of experienced trade negotiators from the Soviet Union. In the past, the official Soviet response to bad harvests had been to decree a general belt-tightening—cattle were slaughtered to reduce the demand for feed grain, and bread and meat were less plentiful until better harvests came. But this time the policy was different; without any public announcements or fanfare, Soviet leaders decided to import grain to make up for the deficit in their own harvest.

The Soviet buyers worked quietly to negotiate deals with several large grain-trading companies to buy a total of 28 million tons of North American grain, 18 million tons of it from the United States. If news of the Soviet negotiations had become public while they were in progress, grain prices probably would have risen significantly. But by working in good free-market style, keeping their transactions private until after the contracts with the trading companies were signed, the Soviet buyers acquired a fourth of the U.S. grain reserves at very favorable prices.

Other hungry nations also came to the United States to buy grain. Altogether, U.S. grain exports increased by 70 percent between mid-1972 and mid-1973, to a record 72 million tons. The result, as Lewis Beman put it in *Fortune* magazine, "was a dramatic increase in farm prices which was quickly translated into an explosion of retail food prices." Grain prices rose sharply, and the consumer price index followed. Retail food prices, which had risen by only 5 percent during all of 1972, shot up 7.8 percent in just the first three months of 1973. According to Beman:

> For all the attention given OPEC's quadrupling of the price of oil, the near trebling of U.S. grain prices was a more immediate cause of the tremendous surge of inflation that nearly overwhelmed the American economy in 1973 and 1974. Transmitted quickly through the complex supply chain that links together what is still the nation's largest industry, the runaway cost of grain pushed the price of bread, meat, and other essentials to heights undreamed of by even the most pessimistic housewife.

How does the climate-crop-consumer machine work? Why should bad weather halfway around the world result in skyrocketing prices in U.S. supermarkets? If the United States had just harvested more than 200 million tons of grain in 1972 and already had 70 million tons on hand, why did exporting 72 million tons have such dramatic effects on the nation's economy? Surely there was no shortage of grain in a country where the government had been paying farmers for years to keep land out of production to avoid crop surpluses.

To answer these questions, and to comprehend the workings of the intricate internal mechanisms of the climate-crop-consumer machine, it is necessary to understand the special role that the North American breadbasket has come to play in supplying food to the rest of the world. And to understand how that role developed, we must look at the tremendous political, social, and economic changes that have taken place over the last 200 years.

The twentieth-century supermarket society of the United States and most other affluent industrial nations, in which the consumer of food is separated from the farmer who raises it by tremendous physical, social, and psychological distances, is a very new phenomenon. Throughout 99.98 percent of the 10,000-year history of agriculture, the vast majority of the world's people subsisted on food that they grew themselves on the land where they lived. The comparatively few people who lived in cities bought their food from farmers who brought their surplus crops short distances to market to obtain cash for a small number of necessities that they could not raise or make on their own land.

But now the situation is very different. In the United States, less than 5 percent of the work force is on the nation's farms; the rest of the people are food consumers, insulated from the producers by a lengthy and complex chain of processing, distribution, and marketing operations.

Among the nations of the world, there are few producers and many consumers. Before World War II, only western Europe imported more grain than it exported; all of the other major regions of the world were net exporters. During the 1934–1938 period, Latin America exported an average of 9 million tons of grain a year, while North America exported 5 million tons. Eastern Europe, including the Soviet Union, also exported 5 million tons per year. Africa's net exports averaged 1 million tons annually, Asia exported 2 million tons, and Australia and New Zealand together exported 3 million tons.

By 1976 this picture had changed dramatically. North American exports that year totaled 94 million tons; Australia and New Zealand exported 8 million tons. All of the other regions were net importers. The North American breadbasket had become the major supplier of the world's food.

How did this shift in the patterns of world grain trade come about? According to agricultural economist Lester Brown, president of the Worldwatch Institute, "If one were to select the single dominant factor transforming world trade patterns in recent decades, it would be varying rates of population growth."

In the early 1950s, Brown points out, North America and Latin America had approximately equal populations—163 and 168 million, respectively. But by the mid-1970s, North America had less than 240 million people, while the population of Latin America had grown to more than 350 million. The same sort of explosive population growth occurred in Africa and Asia. At the same time that population and food demand were increasing in those regions, agricultural technology was pushing North American food production up to unprecedented levels. Reserve stocks of grain accumulated steadily in the United States during the 1960s, and some 50 million acres of U.S. farmland were deliberately held out of production to hold down crop surpluses and keep grain prices up to levels that would keep U.S. farmers in business (Brown, 1975).

The interior prairies and plains of the United States—the breadbasket of the 1970s—were one of the last major agricultural regions of the world to be brought into cultivation. The growth of agriculture in this region occurred during the nineteenth century, a period when agriculture in much of the world was changing. In all of the nations affected by the Industrial Revolution, commercial farming was replacing subsistence farming. Instead of a way of life, farming was becoming a business that required large capital investments and was subject to the vicissitudes of national and world markets.

In colonial days and the early years of the United States, American farmers were independent and largely self-sufficient. In the language of that time, they were husbandmen or yeomen; today we would call them subsistence farmers. Thomas Jefferson's ideal for the United States was an agrarian republic of independent yeoman farmers. Here is how one historian, Dwight Lowell Dumond, described agriculture in the early nineteenth century:

> Farming, for generations, had been a way of life. Every farm homestead had produced food for the family, fuel, and much of the clothing. Houses and barns were built from the surrounding stands of timber. Fuel was cut as needed in the farm wood-lot. Fruit from the orchard was preserved or buried for winter use. Meat was slaughtered in autumn. Eggs, butter, and milk were constantly fresh items in the family diet. Horses and mules, fed from the pasture and corn and oat fields, supplied the power. The few items needed from the outside world were purchased in the rural towns, and a wide variety of products not needed for family use were sold in the local market.

Granted, farm life was not an idyll. Long hours of hard work were required from every member of the family, and there were few opportunities for young people to choose another lifestyle if they found farming uncongenial. But farmers were independent economically; the forces of nature might defeat them, as many American farmers were defeated by the bitter weather of 1816, but the forces of the market were no threat.

Food producers and food consumers were essentially the same people— in 1820, fewer than 8 percent of the people in the United States lived in the cities; the rest lived and worked on or near the farms.

But even as early as the 1830s, this style of American agriculture was changing. In their book *The American Farm*, Maisie Conrat and Richard Conrat point out that, at that time,

> In the immediate vicinity of burgeoning factory towns and commerical centers, farmers were concentrating their energy upon supplying local markets with fresh vegetables, fruits, and dairy products. In outlying districts, farmers had begun to specialize in the raising of beef or pork, or in the production of wool for the region's new textile industry.

By mid-century, as the frontier moved west, farm produce began to move back to the eastern cities by river and canal. After the Civil War, as the railroads boomed, farmers in the interior of the United States shipped their crops east not only for consumption in the cities along the Atlantic coast but for foreign export as well. According to the Conrats, U.S. grain exports rose from 11 million bushels of wheat and 3 million bushels of corn in 1850 to 217 million bushels of wheat and 212 million bushels of corn in 1897. In the years after the Civil War, U.S. grain, meat, and cotton exports to Europe were the mainstay of the nation's economic and industrial growth. The income from these agricultural exports paid the interest on European capital that was used to build U.S. railroads and factories.

With the railroads providing practical transportation between different regions of the United States, agricultural specialization became more widespread. Large-scale wheat production moved westward across the north central states and into Kansas, North and South Dakota, Oklahoma, and Texas. The "corn belt" developed on the midwestern prairies, where the farmers discovered the profitable combination of corn growing and hog raising. During the last half of the nineteenth century, U.S. corn production more than quadrupled, and production of oats and wheat increased seven times.

Technology played a major role in bringing more land into agricultural production. In the early 1800s, good arable land was plentiful in the United States; labor, not land, was the limiting factor for agricultural production. In the 1830s, a farmer and one helper could plant and harvest only about 15 acres of wheat. Then new farm machinery was introduced—cultivators, reapers, threshers, and other labor-saving devices. By the 1890s, two people with farm machinery could plant and harvest more than 250 acres of wheat.

The combination of new machines and expanding agricultural markets accelerated the ecological impact of the traditional American practice of "land-skinning"—farming land intensively until its natural fertility was exhausted, then moving on. As long as there was a plentiful supply of cheap and fertile land further west, many farmers found skinning a profitable practice. Wheat booms—a few years of high production followed by an inevitable decline in fertility—moved west across Wisconsin, then Minnesota, then the Dakotas. It was not until the good land started to give out that agricultural technology was turned from trying to increase the amount of land that each person could farm to learning how to maintain good productivity on existing farmland. Fertilization, crop rotation, erosion control practices such as contour plowing, and other techniques gradually replaced the old practice of farming the land to exhaustion and then moving on.

In nineteenth-century Europe as well as in the United States, the small farmer who lived off the land was being replaced by large-scale, heavily capitalized commercial farming. The rapidly growing urban centers of industry, commerce, and finance created a large consumer class that produced none of its own food. The new technology encouraged productive farming regions to raise large agricultural surpluses for export to other regions or nations, with the result that the producing regions became vulnerable to fluctuations in prices on national and world markets.

Farming as a way of life changed immeasurably in the United States during the nineteenth and early twentieth centuries. Jefferson's agrarian dream of a republic of independent, self-reliant yeoman farmers vanished (see Figure 2). Farm equipment was expensive and required a large capital investment, which the farmer often obtained through a bank loan secured by a mortgage on the land. The livelihood not only of individual farmers and their families but of entire communities in regions where a single cash crop supported the economy hinged on the success of the year's harvest and the state of the market. As the Conrats put it, "For a large segment of American farmers, specialized commercial agriculture failed to bring prosperity. On the contrary, dependence upon cash crops and the fluctuating prices on international markets brought increasing insecurity, hardship and indebtedness."

When European agriculture was disrupted by World War I, wheat prices rose dramatically, and many U.S. farmers went deep into debt to get more land and machinery. There was a short period of high profits, but when the war ended, U.S. wheat exports dropped to half of their wartime level. Though the price of wheat fell, many farmers continued to plant as much wheat as they could in a desperate effort to pay off their

debts. The oversupply depressed the price of wheat even more. Then
came the stock market crash of 1929 and the beginning of the Great
Depression. In 1931, as the depression deepened, the U.S. wheat crop was
the largest ever harvested, and wheat prices dropped lower than ever.
Many workers who had lost their jobs in the cities went back to farming,
where at least they could raise some food for their families. The number
of operating farms in the United States increased from 6.3 million in
1930 to 7.2 million in 1931. But the average annual U.S. farm income was
only $400 per family.

The situation of U.S. agriculture in the early 1930s was a sort of bitter
parody of Jefferson's dream of a society of independent, self-sufficient
yeoman farmers. The free-market system was not working for the
nation's farmers. They had worked hard and competitively and success-
fully to produce more food and fiber for the country and the world, and
now they found themselves beaten down into poverty by powerful and
dispassionate economic forces that they did not understand. The farmers
were not alone in their despair and confusion, of course; by 1932 hundreds of thousands of people were homeless and wandering aimlessly
around the country. Hundreds of thousands of others were living in

FIGURE 2 The picture of agricultural success and prosperity conveyed by this West Virginia farm was a Jeffersonian ideal that began to fade as U.S. agriculture went through great changes in the nineteenth and early twentieth centuries. (United States Department of Agriculture.)

makeshift shelters on vacant city lots in miserable colonies known derisively as Hoovervilles, for the Republican president whom many blamed for the depression.

On March 4, 1933, Franklin Delano Roosevelt became the thirty-second president of the United States and told the people in his inaugural address that "We have nothing to fear but fear itself." Roosevelt backed up his rhetoric with a sweeping program of federal action that made him one of the most admired U.S. presidents in history to some and one of the most hated and scorned to others. Calling Congress into emergency session, Roosevelt sent it a series of bills that thrust the federal government into the middle of the nation's economic, social, and agricultural problems.

The accomplishments of Roosevelt's New Deal administration have been thoroughly documented by many historians, and we shall not attempt to discuss most of them in any detail here. However, one keystone in Roosevelt's recovery program for the nation was the Agricultural Adjustment Act (AAA), designed to reduce farm production and raise farm prices. By the time the AAA became law in June 1933, most of the nation's crops had been planted and new litters of pigs had been farrowed in the corn belt. Secretary of Agriculture Henry A. Wallace immediately ordered the slaughter of hundreds of thousands of baby pigs and the plowing under of large acreages of cotton and other crops. Opponents of Roosevelt's policies declared that this destruction of agricultural products was economic insanity. But most farmers supported the AAA because it promised to raise the prices that they received for their products, even though it demanded that they produce less.

The agricultural policies of the New Deal represented an explicit recognition of the fact that the main problem of U.S. agriculture was that it had succeeded too well in the job of growing more and more crops. From 1933 until 1972, the principal task that would occupy U.S. agricultural policy makers was coping with agricultural overproduction.

Under the AAA, farmers were paid by the federal government to keep part of their land out of production in order to reduce the supply of wheat, cotton, and other crops and thereby raise prices. In 1936, the Supreme Court declared the AAA unconstitutional. But by then drought

FIGURE 3   This abandoned Texas farm, photographed during the worst days of the Dust Bowl era, typifies millions of acres of land that were rendered unfit for agriculture by drought and wind erosion during the 1930s. (United States Department of Agriculture.)

had created the Dust Bowl in the Great Plains; 1935 farm production had been held down as much by the severe and widespread drought as by the restrictions of the AAA (see Figure 3). The damage done to millions of acres of farmland by the dry weather and high winds provided the rationale for a new law, the Soil Conservation and Domestic Allotment Act, under which the government paid farmers to plant part of their land in soil-conserving crops, such as alfalfa and clover, instead of wheat, corn, and other crops that depleted the soil. This law served the dual purpose of preserving and restoring farmland that was threatened or damaged by the drought and, at the same time, limiting commercial agricultural production much as the AAA had done.

Congress passed a new Agricultural Adjustment Act in February 1938. It was designed to control agricultural production and to raise the ratio of farm prices to other prices to "parity"—the level that had existed during the 1909–1914 period. Secretary of Agriculture Wallace established a quota system designed to protect wheat farmers from their own over-competitive tendency to produce too much. Farmers who participated in the plan were assigned acreage allotments, and they voted on whether or not to establish production quotas in every year when a wheat surplus was foreseen. In effect, this program gave farmers the power to control prices and limit production that had long been exercised by industrial corporations. Parity was based on Wallace's concept of an "ever-normal granary," which would adjust supply to demand in such a way that the best interests of both agriculture and industry would be served.

By 1937, agricultural prices had risen more than 80 percent over the disastrous levels of 1932, and farm income was back to about the 1929 level. In addition to restricting agricultural production, the Department of Agriculture had instituted other practices, such as payment of federal subsidies on exported wheat to encourage disposal of agricultural surpluses. But in spite of all of the ingenious New Deal policies for support, farm prices dropped slightly in 1938 and 1939. Even though millions of acres of farmland had been taken out of production, new agricultural technology was continually increasing crop yields. And the farmers, naturally enough, retired their poorest land and kept their most fertile acres in production, increasing the average per-acre yields of the land that remained under cultivation.

Since human beings made the transition from hunting and gathering to tilling and harvesting some 10,000 years ago, there have been at least six major technological innovations that provided great increases in the earth's food-producing capacity. The first was irrigation, which was first used on a large scale about 6000 years ago on the flood plain of the Tigris and Euphrates Rivers. The next major innovation in agriculture was the harnessing of animals to plow the land, which meant that fewer people could till more land.

A third major development was the exchange of crops between the Old and New Worlds. The potato, a New World plant, quickly became a major food crop on the other side of the Atlantic, and today the potato crop in the Soviet Union and Europe is several times larger than that of the New World. Corn, or maize, which was unknown in Europe until it was brought over from America, is a major feed grain in many countries and is raised for human consumption in others. The soybean, which came to North America from China, is now a leading U.S. export crop.

The invention of the internal combustion engine contributed significantly to increased agricultural production. Early farm machinery was drawn by teams of horses that later gave way to cumbersome steam engines. But now a great deal of the world's farmland is worked with machinery powered by flexible, powerful gasoline and diesel engines. The other two major technological innovations in agriculture were the introduction of chemical fertilizers and pesticides and advances in plant genetics that have produced more productive varieties of crops.

Each of these innovations increased food production by permitting more land to be cultivated by fewer people. Not until comparatively recently has more food been produced by raising the yield of each acre of land under cultivation. According to Lester Brown (1974):

> Prior to the twentieth century, rates of increase in output per acre were so low as to be scarcely perceptible within any given generation. Only in recent decades have some countries succeeded in achieving an increase in output per acre sufficiently sustained and rapid to be considered a yield "takeoff."

In the 1930s, agricultural technology continued to improve the potential productivity of U.S. agriculture, even while land was being withdrawn from production to reduce crop surpluses. During the decade of the 1930s, about 160 million acres—more than 15 percent of the land that was being farmed in 1930—was taken out of agricultural production. But during that same decade, new high-yield hybrid corn was being planted in more and more of the U.S. corn belt. Combines, which harvested and threshed wheat and other small grains in a single operation, were replacing older and less efficient machines that harvested and threshed separately. Airplanes were being used for crop dusting, and new chemicals were becoming available for use against plant and animal diseases.

There were several reasons for this paradox of continued development and adoption of new technology for increasing agricultural production during a period when massive national programs were underway to prevent overproduction. One was simply the enthusiasm that new inventions and techniques had traditionally inspired in the United States. During the late nineteenth century and the early years of this century, the nation's natural resources seemed inexhaustible, and economic and industrial growth was an overriding national goal. Thus the general attitude toward new technology was positive—if we *can* do it, we *should* do it.

The farmer's eagerness to adopt new technology grew, to a great extent, from the same characteristics of U.S. agriculture that had caused it to produce itself into poverty in the years after World War I. In the language of the economist, farming is a competitive economic activity; that is, the number of agricultural producers is so large that the production of individual farmers does not affect the price that they receive for the crops that they produce. In the short run, farmers who can produce twice as many bushels per acre as other farmers will make twice as much money. It is not until many farmers double their yield that the surplus supply of the commodity may force the market price down. Thus the ambitious farmer will try to be among the first to adopt new technology in order to get ahead of the competition. But if the times are not right for it, this kind of free-market competition can be destructive to the agricultural community as a whole.

In a period when food demand is both growing and elastic, enthusiasm for new technology that will increase agricultural production is beneficial to both the farmer and the consumer. As Earl Heady of Iowa State University pointed out in *Scientific American* in 1976:

> Although [food demand was elastic] in the U.S. until early in the 20th century, by the 1920's the per capita income had risen high enough for the domestic demand for food to have become highly inelastic. That is, incomes had risen to the level where consumers were able to buy all the food they needed and further increases in income could have little effect on food consumption. The result of such inelasticity is that an increase in food output of 1 percent leads to a decrease in food prices of more than 1 percent. If other factors, such as exports, remain at constant levels, an increase in farm production greater than the rate of population growth causes the market price of food to decline at a rate greater than the growth of demand.

This sort of decline occurred in the United States in the years after World War I and on into the early 1930s. It was reversed by the farm policies of the Roosevelt administration, but by 1939 farm prices had declined slightly from the pre-Depression levels to which they had risen in 1937. Then Hitler invaded Poland, plunging the world into another war that would bring terror and misery to hundreds of millions of people but new prosperity to the U.S. farmer. In 1940, as England, France, and other Western European countries found themselves fighting Germany instead of raising food, they turned to the United States as they had during World War I. And this time, although there were no new frontier lands to plow, U.S. farmers were able to apply the new agricultural technology to produce larger yields from each acre of farmland, including most of the land that had been left fallow or planted in cover crops since the depression and Dust Bowl years.

A decade of prosperity followed for U.S. agriculture, as the wartime demand for food was followed by heavy exports to liberated and occupied territories whose agricultural production had been devasted by the war. By 1950, however, the situation of the 1920s was returning. Demand for U.S. agricultural commodities, both at home and abroad, was inelastic. Farm production continued to rise, and prices for grain and other farm commodities declined. The consumer benefited as retail food prices dropped, but the farmer was getting deeper and deeper into trouble. Another war, this one in Korea, absorbed some of the U.S. agricultural surplus, but when it ended in 1953, prices for agricultural commodities started sliding downward again.

At the same time that demand for U.S. agricultural products was dropping off, new agricultural technology was pushing U.S. farm production upward. In the 1950s, U.S. corn yields began rising in a trend that would more than double the number of bushels of corn produced by each acre of land over the next two decades. Yields of wheat and other grains also increased substantially, if not as dramatically as corn yields. Heavy use of nitrogen fertilizer was a major factor in increasing yields, and chemical herbicides and pesticides and other technological innovations also contributed significantly.

As before, the biggest problem of U.S. agriculture was its success at raising crops. In 1860, each farm worker produced enough food and fiber to supply about four people. By 1900, each farm worker was supplying seven people, but by 1950 each was raising enough for fifteen. The competitive economic activity of U.S. farmers had resulted in steadily increasing production on each farm and ruinous overproduction by the U.S. agricultural establishment as a whole in terms of actual free-market demand.

One approach that the federal government took to agricultural surpluses in the 1950s was to try more of the same remedies that had worked in the 1930s. Farmers were paid to keep land out of production, and the government bought surplus commodities and stored them. But in 1954, the United States Congress passed a piece of legislation that approached the problem of U.S. agricultural surpluses in an entirely new way— Public Law 480 of the Eighty-Third Congress, the Food For Peace Act.

One writer on food and foreign policy, Emma Rothschild (1977), has described P.L. 480 as "a law which has changed the diet and the political life of half the world." That description is not unduly extravagant; it is P.L. 480, along with the differential rates of population growth of the United States and the regions of the world that formerly exported grain and now have to import large amounts to feed their people, that account for the dominant role that has fallen to the North American breadbasket as supplier of food to the rest of the world.

According to Rothschild, P.L. 480 had four purposes: to get rid of surplus food, to help hungry people, to further U.S. political and military policy, and to create markets for American food. Title I of the law authorized the United States to make long-term, low-interest loans to other countries to buy U.S. farm products. Title II provided for outright donation of U.S. food either to foreign governments or to international relief organizations such as CARE. It is important to note that, although uncounted millions of hungry people have been fed under the provisions of P.L. 480, its humanitarian purposes have nearly always taken second place to its role as an instrument for disposing of surplus crops. During the world food crisis of 1973–1974, after mountains of U.S. grain had been sold to the Soviet Union and other countries that were able to pay, U.S. food aid to many poor countries was cut back. When famine struck Bangladesh in 1974, according to Rothschild, that country received less food from the United States than it had in earlier years when disposal of U.S. surpluses was considered a major goal of the Food for Peace program. Moreover, of the 80 countries receiving some food aid from the United States in the mid-1970s, some were quite prosperous and others were on the list for reasons that were closely tied to military policy. In 1972, for example, South Korea was the leading recipient of U.S. food aid; in 1973 and 1974 South Vietnam headed the list; in 1975, the largest recipient was India; and in 1976 Israel received more than any other country. Other countries, such as those in the drought-stricken Sahel, received large shipments of food as short-term emergency aid, of course (Figure 4).

Perhaps the main cause of both the successes and the failures of P.L. 480 can be found in its tendency to try to be all things to all people. As Rothschild put it: "Some congressmen saw it as aid to Kansas, and others as aid to Asia."

As aid to Kansas—providing a safety valve for U.S. agricultural surpluses—the Food for Peace program was highly successful. In the mid-1960s, up to 18 million tons of food a year was going overseas under P.L. 480. But even these massive food shipments could not keep up with population growth in many hungry nations. In his 1967 State of the Union Message, President Lyndon Johnson said, "Next to the pursuit of peace, the really greatest challenge to the human family is the race between food supply and population increase. That race tonight is being lost."

In a report issued the following May by a panel of the President's Science Advisory Committee, the world food problem was summarized as follows:

> Despite expenditures of billions of dollars for foreign aid; despite donations and concessional sales of millions of tons of food to developing nations; despite herculean efforts by numerous voluntary groups; despite examples of highly productive technical assistance programs by foundations; and despite years of activity by international organizations . . . there are more hungry mouths in the world today than ever before in history.

Both the Kennedy and Johnson administrations attempted to promote economic development in the hungry nations, and after 1966 countries that received food aid under Title I of P.L. 480 were required to improve their own agricultural programs. However, most of the developing nations that received large quantities of food from the United States became increasingly dependent upon it. This was not entirely inadvertent; it was another aspect of the mixed motives that had created the Food for Peace program and kept it alive. In the cold-war days of the 1950s, food aid was seen as a nonmilitant way to help friendly nations build military establishments by allowing them to divert resources from agricultural production. And U.S. food aid also encouraged the growing predilection of new Asian and African nations to neglect their own agriculture in favor of glamorous development projects like national airlines, dams, and urban construction.

The pessimistic view of the world food situation expressed in 1967 by President Johnson and his advisory panel gave way in the last years of the decade to a growing confidence that agricultural technology could continue to expand yields of food crops sufficiently to feed the world. Between the mid-1960s and 1971, total world food production grew more than 25 percent, largely because of the Green Revolution—the use of new high-yield varieties of wheat and rice with fertilizer, other chemicals, irrigation, and improved cultivation techniques. This "package" of agricultural technology has produced dramatic increases in crop yields in some of the countries where a high rate of population growth and low agricultural production have gone hand in hand in the past.

The Green Revolution began in Mexico in 1943 when the Mexican government and the Rockefeller Foundation established an agricultural research project with a highly pragmatic goal: to increase Mexican wheat production as much as possible in as short a time as possible. This project brought together a team of agricultural specialists in plant breeding, plant pathology, soils, entomology, and farm management, under the direction of Norman Borlaug (Figure 5).

One of the first problems that they attacked involved the response of traditional wheat varieties to increased amounts of fertilizer. When more than 40 pounds of nitrogen fertilizer per acre was applied to the wheat, it grew tall and top-heavy. Severe crop losses often resulted from "lodging"—bending of the plant to the ground under its own weight. Borlaug and his colleagues developed dwarf varieties of wheat with short, stiff

FIGURE 5 Dr. Norman E. Borlaug, who received the 1970 Nobel Peace Prize for his role in the Green Revolution, cross-pollinating wheat at the International Maize and Wheat Improvement Center (CIMMYT) in Mexico. (The Rockefeller Foundation.)

stems that would give greatly increased yields with applications of as much as 120 pounds of nitrogen fertilizer per acre. These new high-yielding varieties were also bred for reduced sensitivity to the length of the day, permitting them to be planted at diverse latitudes and times of year. By applying the right amounts of fertilizer, water, and pesticides to these new high-yielding varieties (HYVs) of wheat and using the right techniques of cultivation, the Rockefeller team achieved remarkable increases in yields. By the mid-1950s, Mexican wheat yields had doubled, and by the late 1960s they had doubled again.

Encouraged by the success of the Mexican wheat research, the Rockefeller and Ford Foundations established the International Rice Research Institute in the Philippines in 1960. Again, an interdisciplinary team was brought together, this time under the direction of Robert Chandler. They developed a "miracle rice" known as IR-8, which could double rice yields in many parts of Asia with the proper growing conditions and cultivation techniques (Figure 6).

At first, as the new high-yielding varieties of wheat and rice were

FIGURE 6 A worker at the International Rice Research Institute in the Philippines removes the anthers of a rice panicle in preparation for crossing two varieties of rice. (The Rockefeller Foundation.)

adopted in many Asian and North African countries, they seemed to some people to have solved the world food problem. India, Pakistan, Turkey, the Philippines, Indonesia, Malaysia, and Sri Lanka all achieved great increases in grain production. But the experience of Mexico, the first country to benefit from the Green Revolution, clearly indicated that the Green Revolution was no panacea for the problem of increasing food demand generated by runaway population growth. In the 1940s, Mexico was importing half of the wheat that its people consumed. By the late 1950s, Mexican wheat yields had doubled, and the country was self-sufficient in wheat even though its population had grown substantially. By the late 1960s, wheat yields had doubled again, and Mexico had become a net exporter of grain. But by the early 1970s, with one of the world's highest rates of population growth, Mexico was importing food again to feed its people. The remarkable gains in agricultural production achieved by the Green Revolution had not solved Mexico's food problem; the real problem appeared to be too many people for the nation's agricultural resources.

In recent years, the widespread enthusiasm that first greeted the Green Revolution has given way to skepticism in some quarters. Critics have maintained that, unless they are given generous amounts of fertilizer and water and are cultivated in just the right way, the HYVs give lower yields than the old traditional varieties of wheat and rice, and their yields drop by a higher percentage under unfavorable climatic conditions. The people who developed the new varieties deny this allegation, however. Keith Finlay, deputy director of the International Maize and Wheat Improvement Center, usually called CIMMYT (the acronym for its name in Spanish), says:

> In general, the new varieties are more productive over the full range, from no fertilizer or water to the ideal conditions we have at CIMMYT. The reason is that a dwarf plant is more efficient: it puts more of its energy and total plant matter into grain than a taller plant does. Many countries can and do grow these new varieties without fertilizer, and they still get yields as good as or better than their native varieties. (Anderson, 1975)

Although the Green Revolution has not lived up to the hopes that some nations and people built up for it, it almost certainly saved a great many people from starvation and malnutrition. It also bought some time for nations with rapid population growth to try to deal with this problem, although the time was not used very effectively. As Lester Brown points out (1975):

> These governments could use the breathing space afforded by the new production gains to begin to bring population growth under control, or they could defer making the difficult policy changes until they were again confronted with crises. Unfortunately, all too many countries have followed the latter path.

Many of the countries to which the Green Revolution brought the most dramatic crop yields, such as Mexico, are again dependent on the North American breadbasket for grain to feed their increasing numbers of hungry people. Population growth in other countries has pushed the United States rapidly toward one of the goals of P.L. 480—creating expanding foreign markets for U.S. agricultural production. Food aid under P.L. 480 reached a high of about 18 million tons in the mid-1960s, then started down, dropping to 11 million tons in 1970 and 3.3 million tons in 1974. But as aid went down, commercial exports went up. In 1972, food exports financed by the Food for Peace program amounted to $1.1 billion, while commercial exports were worth $1.7 billion. By 1975, the Food for Peace exports still totaled $1.1 billion, but the commercial exports had increased fourfold, to $6.8 billion. Although a good part of this shift in export patterns obviously is attributable to population growth, increasing affluence, and unfavorable weather in other countries, there is little doubt that P.L. 480 played a major role in creating the overwhelming dependence on the North American breadbasket that characterized world grain-trade patterns by the mid-1970s.

Like U.S. agricultural production itself, the Food for Peace program has, in a sense, created serious problems by succeeding too well at what it set out to do. When the agricultural surpluses of the 1950s and 1960s disappeared, so did the moderate and stable retail food prices of that era. As Lester Brown wrote in 1975:

> Today the entire world is living hand to mouth, trying to make it from one harvest to the next. Global food insecurity is greater now than at any time since the years immediately following World War II, making weather in the principal food producing countries a major global economic and political concern.

Many people in the U.S. agricultural establishment considered Brown an alarmist, especially after bumper crops of grain were harvested in the United States and other countries in 1975 and 1976. But others pointed out that, as long as the United States exports its grain commercially, selling it at the free-market price, large U.S. grain stocks do little to feed people in poor countries that cannot compete with customers like Japan and the Soviet Union.

Many scientists and other people saw climatic variability as a serious and continuing threat to world food supplies. Although they were derided as prophets of doom by some agricultural policy makers, by 1977 a number of climatologists, agricultural scientists, and researchers from other disciplines had developed a carefully documented case for the potential vulnerability of agriculture to climatic change and fluctuation.

One result of the climatic and economic anomalies of 1972 had been a sudden upsurge of public interst in climatology, which had long been regarded as the dullest branch of atmospheric science, concerned mainly with calculating averages of long-term weather records. By 1974, which turned out to be another bad crop year in many parts of the world, this interest grew even more intense. As Francis Bretherton, director of the National Center for Atmospheric Research (NCAR) put it, "Climate research is emerging as the Cinderella of meteorology, after having been rejected as a hopeless endeavor for many years by all but a handful of dedicated statisticians."

Climatologists who had been making pronouncements about climatic trends for years without receiving much attention found themselves speaking to attentive audiences of policy and decision makers at Congressional hearings and White House briefings. A spirited national and international debate began over the nature of current climatic trends and their potential impacts on food production. A number of distinguished climatologists, including Reid Bryson of the University of Wisconsin, Hermann Flohn of the University of Bonn, and Hubert Lamb of the University of East Anglia, suspected that the global cooling trend that had begun in the 1950s might continue for some time and that the cooling might be accompanied by increasingly variable weather. Both of these developments would have serious implications for agriculture.

Some other atmospheric scientists with equally sound credentials, such as B. J. Mason, head of the British Meteorological Office, J. Murray Mitchell of the U.S. National Oceanic and Atmospheric Administration, and Helmut Landsberg of the University of Maryland, saw no convincing evidence that any major climatic change was in progress.

"The world does seem to have been cooling off in the past 20 years or so," Mitchell said, "but we shouldn't count on this cooling trend—ominous or not—continuing for a long time in the future. There is no way we can tell whether it will or not."

And Landsberg maintained that "atmospheric conditions as a whole, while wildly oscillating from year to year, have a fair amount of stability. They pendulate around averages that change only slowly."

But even these climatological conservatives saw a danger that global food production could be seriously affected by climatic fluctuations in important agricultural regions. In testimony before a Congressional subcommittee, Landsberg said that "the weather in [food] producing areas of the world will stay highly variable, and the only insurance is to have an adequate reserve on hand to compensate for the inevitable occasional crop losses." And in a letter to the administrator of NOAA, Landsberg proposed a study of the potential impacts of weather and climate on U.S. crop production.

Some agricultural scientists and many USDA officials maintained that the climatologists were sounding false alarms and that their fears were groundless. According to Louis Thompson, Associate Dean of agriculture at Iowa State University, "There was frequent reference in the early 1970s to the fact that technology had advanced to such a level that weather was no longer a significant factor in grain production." Thompson, who disagreed emphatically with this position, was one of a group of atmospheric and agricultural scientists who were invited to participate in the study that Landsberg had proposed to NOAA. In 1973, NOAA published a report of this group entitled *The Influence of Weather and*

Climate on U.S. Grain Yields: Bumper Crops or Droughts. Among the conclusions of this report were:

- Normal weather in the grain growing regions results in high yields. Whenever the weather departs from normal, it is much more likely that lower yields will result.

- In recent years there has been a remarkable run of near-normal weather—or even that relatively unusual weather that produces even higher yields. The reliability of grain yields in recent years is due to an extraordinary sequence of favorable growing seasons. This cannot be expected to continue.

Responding to this report, a USDA committee said:

> A comprehensive study published by this Department in 1965 evaluating the effect of weather and technology on corn yield in the Corn Belt for the years 1929 through 1962 concluded that through the use of better varieties and improved cultivation and fertilization practices, man has reduced variation in yields in both good and bad weather.

In May 1974, an international group of specialists in such diverse fields as climatology, agricultural economics, geography, oceanography, law, and political science met at the University of Bonn for a workshop on "The Impact on Man of Climatic Change," held by the International Federation of Institutes for Advanced Study. They concluded that:

> The nature of climatic change is such that even the most optimistic experts assign a substantial probability of major crop failures within a decade. If national and international policies do not take such failures into account, they may result in mass deaths by starvation and perhaps in anarchy and violence that could exact a still more terrible toll.

In June 1974, Reid Bryson of the University of Wisconsin and Stephen Schneider of the National Center for Atmospheric Research were invited to Washington for informal discussions with a group of White House policy makers and other federal officials about the implications of climatic variation for food production. According to Schneider, he and Bryson argued that government complacency toward climate-food problems could lead to food-price inflation in well-fed countries and widespread starvation in poor lands. But they did not convince everyone who was there. As Schneider tells it (1976):

> Technology has reduced the dependence of crop yields on weather, I was told by a senior official of the U.S. Department of Agriculture . . . and there is no

real threat of famine (and, by implication, little need for building food reserves). We have an agricultural policy, he said, a policy of plenty.

James McQuigg, a NOAA climatologist who worked with Louis Thompson and others in the study that Landsberg had suggested, was convinced that the complacency of people like this USDA official was based on their failure to recognize that the period from the late 1950s through the early 1970s was one of unusually good growing weather in the North American breadbasket. Other climatologists, including conservatives like Murray Mitchell who did not accept the cooling trend of that period as a long-term climatic change, nevertheless agreed with McQuigg and Thompson that it was a time of unusually benevolent weather for agriculture. Early in 1975, Mitchell said in an interview:

> From the agricultural productivity point of view, the climate's not going to get better. It can only get worse. It could be more drought, or more premature autumn cold, weather-related problems in the spring, or any combination of those things. . . . If there's anything we can be reasonably confident about in terms of projections of future climate, it is that the climate of our crop-growing areas will become more variable than it has been in the recent past.

NOAA was not the only government agency that harbored skepticism about the USDA's complacency toward the impacts of weather and climate on crops. In the summer of 1974, the Central Intelligence Agency (CIA) published an unclassified report entitled *Potential Implications of Trends in World Population, Food Production, and Climate.* In a section entitled "Key Judgments," this report concluded that:

> Trying to provide adequate world food supplies will become a problem of over-riding priority in the years and decades immediately ahead—and a key role in any successful effort must fall to the US. Even in the most favorable circumstances predictable, with increased devotion of scarce resources and technical expertise, the outcome will be doubtful; in the event of adverse changes in climate, the outcome can only be grave.

The CIA concluded that, if climate change caused serious food shortages in other countries, "there would be increasingly desperate attempts on the part of powerful but hungry nations to get grain any way they could. Massive migrations, sometimes backed by force, would become a live issue and political and economic instability would be widespread."

How was it possible for some policy makers at high levels of government to maintain that U.S. crop yields were practically independent of weather and climate conditions, in spite of evidence to the contrary that convinced not only climatologists but intelligence specialists as well? Why couldn't the issue be settled simply by comparing crop yields in years when the weather was good with the yields for bad weather years?

The main problem with this approach lies in the fact that, during the years when, according to McQuigg and his colleagues, the world had a long run of good growing weather, agricultural technology was advancing at an unprecedented rate. Separating the increases in yields attributable to good weather from those achieved through plant breeding, improved farming practices, and increasing fertilization is not an easy task. In 1972, for example, when U.S. grain production dropped below the 1971 level because of bad weather, the yield was still higher than it had been for any year prior to 1971. If the effects of weather on crop yields were to be discerned, it would have to be by comparisons of relative rates of increase rather than absolute yields for the years being considered.

Some impacts of weather and climate on crop yields are dramatically obvious, of course, and cannot be denied by even the most optimistic agricultural technologist. When young winter wheat is literally blown out of the parched ground by high winds, as it was in the Dust Bowl days and again in the mid-1970s in some parts of the high plains, it would be difficult to maintain that the wheat crop was not affected by weather. When a large part of the Florida vegetable crop was killed by freezing temperatures in the Winter of '77, the role of weather was undeniable. And good weather, or at least the absence of bad weather, is usually given some credit for contributing to bumper crops such as the record U.S. corn and wheat harvests of 1975 and 1976.

The debate of the early 1970s was not focused on extremes like these, but was essentially over whether or not U.S. crop yields could be predicted by projecting into the future the rates of increase that had been achieved in the 1960s. Many USDA policy makers felt that this was a valid technique for forecasting future yields; weather and climate were random variables, they said, and could not be taken into account in any systematic way. But scientists like Thompson and McQuigg maintained that, even if climate forecasting could not be done with any precision, a sort of actuarial forecast indicated strongly that extended periods of unfavorable weather for agriculture had occurred in the past in the United States, and there was a high probability that they would occur again.

By the mid-1970s, the latter view was supported by an increasing number of scientific studies that had been undertaken in response to the obvious impacts on food production of the bad weather of 1972 and 1974.

The 1973 NOAA report by McQuigg, Thompson, and their colleagues was not based on conjecture. Its conclusions grew out of studies with a model that correlated actual year-to-year variations in corn, wheat, and soybean yields for the five principal U.S. grain-growing states between 1890 and 1970 with fluctuations in temperature and precipitation during that same period.

Thompson then used the computer model to project the agricultural technology of 1973 backward through the 1890–1970 period, simulating the yields that would have been produced each year if 1973 grain varieties, fertilization, cultivation practices, and other modern agricultural technology had been available. By doing this, he sorted out the effects of technology from those of weather, and in effect compared the year-by-year yields as if there had been no changes in agricultural technology. This model clearly showed low yields of corn during the Dust Bowl years of the 1930s as well as reduced yields during other drought periods. It also indicated high yields and low variability in yields during the period from the late 1950s to the early 1970s. Since the model in effect removed the influence of changing technology on yields and yield variability by holding it constant at the 1973 level, McQuigg and Thompson maintained that "there is strong evidence that it is, indeed, 'good' weather, as well as recent technological gains, that has produced these high yields" and that the evidence indicated that "technology has not influenced the susceptibility of crop yields to weather." The report concluded: "The message is strong that we have been unusually fortunate in recent years to have experienced such high grain yields. It is imperative that we not be lulled into a dangerous and unjustified expectation that such fortunate circumstances will continue" (National Oceanic and Atmospheric Administration, 1973).

Other studies supported these conclusions. In late 1975, The Institute of Ecology and the Charles F. Kettering Foundation organized a study of *The Impact of Climatic Fluctuation on North American Food Crops*. The panel that conducted this study was chaired by James Newman of the Agronomy Department at Purdue University, and it included specialists in agricultural science and other scientific disciplines from several universities and other organizations. The panel decided to demonstrate the impact of climatic variations on crop yields by constructing scenarios in which climatic conditions from four past periods during the last 100 years would be imposed on North American wheat, corn, soybean, and sorghum production with 1975 crop acreage and 1973 technology. Thompson's model was used to select periods of several years when agricultural production was severely reduced by climatic stress (1933–1936), moderately reduced by climatic stress (1953–1955), very high because of favorable weather (1961–1963), and highly variable because of year-to-year climatic variability (1971–1975).

When the scenarios were run by computer, the results indicated that, to put it simply, North American crop yields are still highly susceptible to climatic influences. Here is how the conclusions were summarized in the study report:

> Food production could be greatly reduced by a recurrence of 1933–36 weather conditions. For example, if the 1936 weather recurred, the United States could lose 71 million metric tons equivalent to 27% of the 1975 production for corn, wheat, sorghum, and soybeans combined. Less severe reductions could occur in the 1953–55 scenario. The 1961–63 scenario represented the highest production levels, and the 1971–75 climate could cause production to vary widely from year to year.

Other studies have produced less dramatic conclusions that may have more significance in terms of projecting crop yields over periods of many years in the absence of specific forecasts of climatic conditions. For example, most of the U.S. corn belt normally receives less than 8 inches of rain during the critical period from early June through August. In wet years, when the rainfall reaches 10 or 11 inches in these months, the yield is much higher than in the normal years. Similarly, normal maximum summer temperatures are too high for optimum yields in many U.S. wheat-growing regions. Summers that are slightly cooler than normal result in appreciable increases in wheat yields (Figure 7).

In addition to their direct effects on crop yields, climatic fluctuations can have significant effects on pests that attack crops. Occasionally these effects can be favorable; one modest benefit of the bitter cold of the Winter of '77 in the corn belt was a probable reduction in some agricultural pests. According to James Newman of Purdue University, the low temperatures probably killed off some insect pests such as corn borers and rootworms, as well as diseases, such as corn blights, that overwinter on crop refuse.

But the effects of weather and climate anomalies on insect pests and plant diseases can also be devastating. A classic example is the 1848 potato famine in Ireland. Unusually wet weather in 1845 and 1846 was followed by torrential rains in 1848. The excessive rainfall alone would have reduced the potato harvest to some extent, but the wet conditions also favored the growth of a fungus blight that attacked the potatoes. The blight caused a total crop failure. A million and a half people died, and a great wave of Irish immigrants came to the United States to escape starvation.

During the North American Dust Bowl years of the 1930s, there were

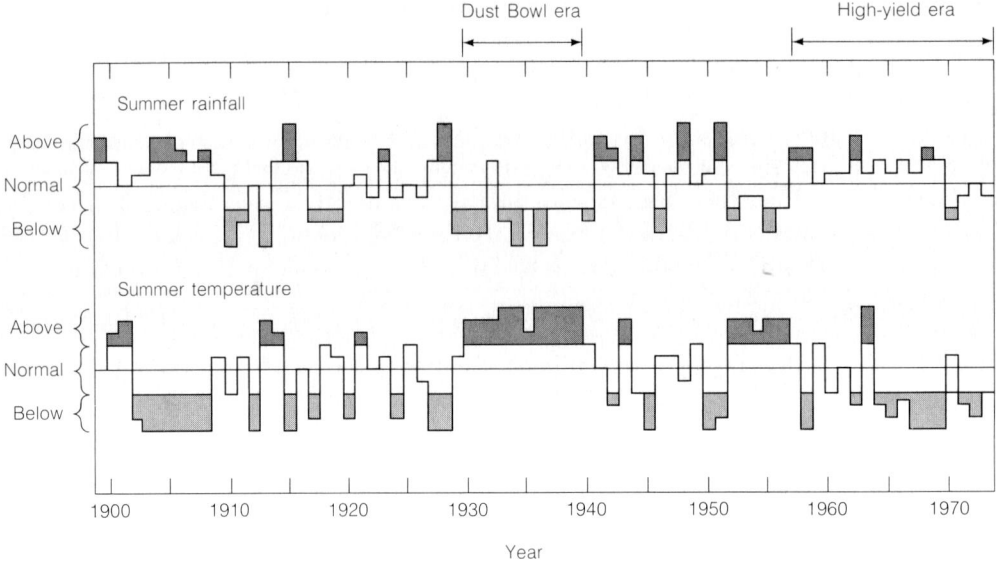

FIGURE 7 This graph of rainfall and temperature in five wheat-growing states (Oklahoma, Kansas, Nebraska, South Dakota, and North Dakota) between 1900 and the early 1970s, was compiled by Dr. Donald Gilman, chief of the Long Range Prediction Branch of the National Weather Service. It shows the relationship between these two critical aspects of climate and the yields of wheat in the region. (National Weather Service.)

high crop losses to some pests that were encouraged by the hot, dry weather. Some of them are described in a report entitled *Climate and Food* published in 1976 by the National Academy of Sciences:

> The pale western cutworm, *Agrotis orthogonia*, took advantage of the change in climate. During May, when its larvae are most active, less rain and warmer temperatures favor outbreaks. Hence, this pest became abundant and destructive in thousands of acres of wheat in the Canadian Prairies and Montana. In one survey in 1937 in Alberta the losses ranged from 10 percent to total destruction of the crop. The cutworms retreated, and considerably less damage was inflicted on wheat after rain returned in the 1940s. . . .
>
> While the cutworm spread in the North, the prickly pear cactus (*Opuntia* sp.) headed eastward out of Colorado. During the drought, this thorny pest invaded . . . pastureland in western Kansas. When the rains came at last in the 1940s, changing the weather, the cactus weed retreated westward to its former home.

The report also describes how stem rust fungus, which has always done serious damage to wheat, caused losses of millions of tons of wheat during the Dust Bowl years. "The evidence of history," the report says, "is that the hot weather of the 1930s favored rust while drought hindered the wheat. The result was disaster."

Even in recent years, with chemical pesticides and fungicides and other tools of modern agricultural technology available, weather-induced outbreaks of pests can still have serious impacts on crop yields. In 1970, a wet spring in the southeastern United States was followed by warm, wet summer weather in the corn belt. A corn leaf blight fungus appeared and spread rapidly, producing shriveled leaves, decayed ears, and rotting stalks across southern cornfields. By mid-August, the blight had spread north through Illinois and Indiana and over a large part of the corn belt. The total loss was estimated at 1 billion bushels—15 to 20 percent of the U.S. corn crop.

The 1970 corn leaf blight is a particularly interesting example of the effects of weather on crop pests and crop yields. It occurred not only in spite of technology but, to a great extent, *because of* modern agricultural technology. In the 1950s, plant geneticists developed a new technique for producing seed corn for high-yield hybrid varieties of corn that, over the last several decades, had replaced the old traditional varieties that gave much lower yields (Figure 8). This new technique, which incorporated male sterility into one parent plant, was more efficient and effective than the old methods that involved physical removal of the tassels from corn plants that were to produce the hybrid seed. But it resulted in nearly all of the seed corn used in the United States being genetically identical, descended from a single plant selected in Texas by the plant breeders around 1950. This meant that nearly all of the corn grown in the nation shared a hereditary susceptibility to certain diseases. In 1970, one of those diseases appeared: a particular variety of a fungus called *Helminthosporium maydis*, which was first identified in the corn belt in 1969 and was later discovered to have been present even earlier. This fungus, which the agricultural scientists called Race T because it was particularly virulent on the widely used variety of corn that originated in Texas, thrives under warm, damp conditions. Thus in 1970, the stage was set: Most of the U.S. corn crop was highly susceptible to the Race T blight, and the fungus was present in U.S. corn-growing areas. When the weather turned warm and wet, the blight exploded across the nation's cornfields, wiping out the highly susceptible plants acre by acre until a billion bushels of corn had been destroyed. But in 1971, farmers substituted a different variety of corn, and the Race T fungus was no longer a serious threat.

Another kind of secondary effect of weather and climate on crops that is sometimes overlooked has its roots in human behavior and agricultural economics on the scale of the individual farmer. Louis Thompson of Iowa

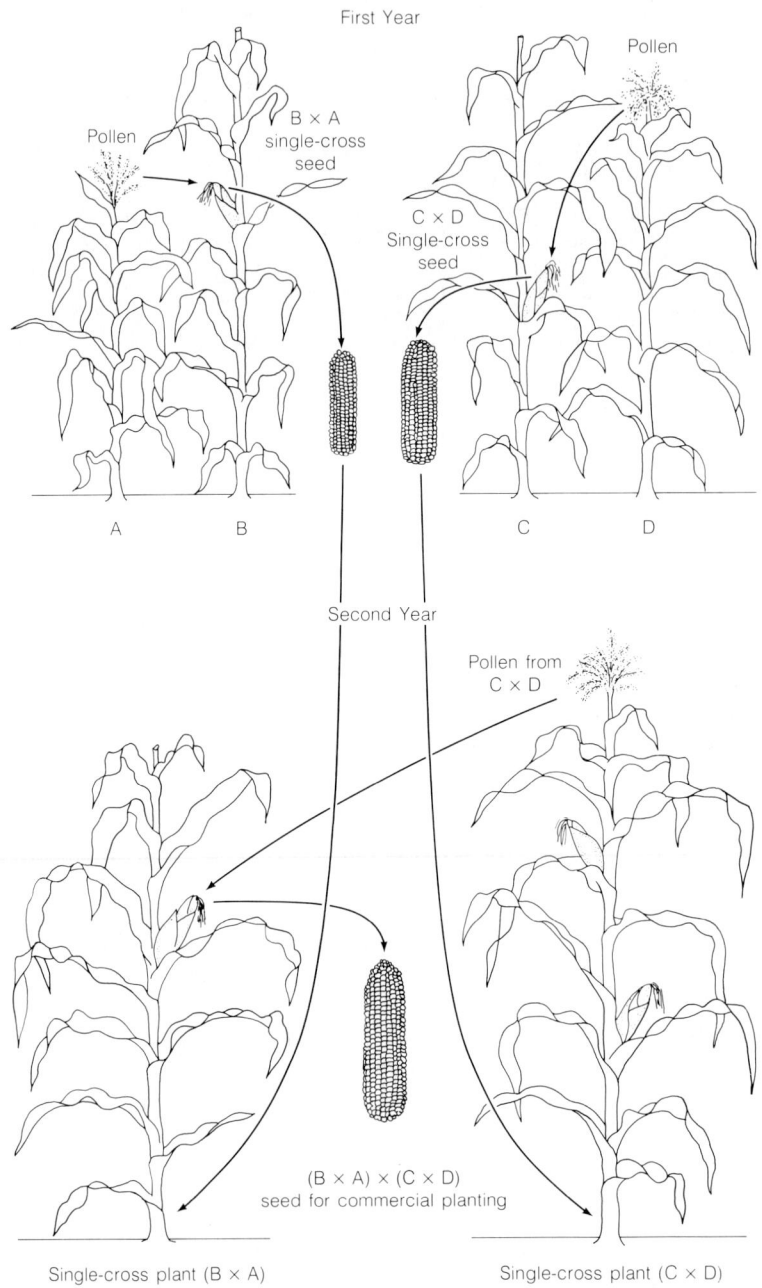

First Year

Pollen

B × A
single-cross
seed

Pollen

C × D
Single-cross
seed

A   B   C   D

Second Year

Pollen from
C × D

(B × A) × (C × D)
seed for commercial planting

Single-cross plant (B × A)   Single-cross plant (C × D)

FIGURE 8 Double-cross method of producing hybrid corn seed. The inbreds shown at the top are crossed to produce seed to give the single-cross hybrids shown below. The single-cross plants are then crossed to produce seed to be sold to the farmer. (From "Hybrid Corn" by Paul C. Mangelsdorf. Copyright © 1971 by Scientific American, Inc. All rights reserved.)

FIGURE 9 Wheat is inoculated with stem rust to test its genetic resistance at the International Maize and Wheat Improvement Center (CIMMYT) in Mexico. (The Rockefeller Foundation.)

State University points out that the consistently good growing weather of the 1960s in the United States contributed to high crop yields in two ways. First, it directly influenced the crops by providing good growing conditions. Good crops also mean good profits for the farmer, who then has more capital available to invest in next year's crop. As Thompson puts it:

> When there's a bad year, the farmer pulls back on his investment. When crops are consistently good for several years, he will spend more for good seed, fertilizer, and chemicals to control insects and diseases. So the consistently good weather of the 1960s contributed to high yields both directly and indirectly.

Thompson also sees the so-called cattle cycle as a subsystem of the climate-crop-consumer machine. Sometimes called the corn-cattle cycle, this delayed response of the livestock industry and retail meat prices to grain prices has been well known to economists for years, although it has not generally been linked with climatic fluctuations.

When the price of corn goes up, as it did after drought hit the U.S. corn belt in 1974, the farmers and grain dealers who have corn to sell profit immediately. But the livestock feeders, who operate feed lots where young cattle are fattened for market on a rich diet of grain and high-protein feed supplements, immediately start losing money as their production costs go up. As their profit margin drops, the first reaction of most cattle feeders is to increase the rate of slaughter, selling off more cattle rather than continuing to feed them expensive corn. This increases the supply of beef, and the price drops, aggravating the cost-versus-profits squeeze in the livestock feeding industry, which usually responds by increasing the rate of liquidation of its herds.

For the consumer, the first effect of this process is favorable: As the market becomes glutted with beef, the retail price drops. But once the livestock feeders have reduced their herds to a low level, the supply of beef drops below the demand for it, and the price of steak in the super-markets starts rising. If, in the meantime, corn prices have dropped, the increasing beef prices encourage the cattle feeders, who start holding back heifers to build their herds up again. This reduces the supply of beef even more, the price keeps rising, and the cattleraisers hold back even more cattle, hoping for better prices. This trend usually continues until corn prices rise, starting the whole cycle over again.

The cattle cycle usually takes about ten years to run its course. There have been seven of these cycles since the turn of the century, according to Richard McDougal of the American National Cattlemen's Association, with a build-up of herds for five to eight years and a liquidation phase of three or four years. In early 1977, the reduction of herds that began in 1974 appeared to be nearing its end, with a new build-up period and high retail beef prices ahead. Similar cycles exist in hog and chicken production. As the length of the cycle depends on the amount of time required to bring the animals to market weight, the corn-hog cycle is not as long as the cattle cycle; the chicken cycle is even shorter.

Traditionally, the cattle cycle has been regarded as an artifact of the free-market economy, governed by the law of supply and demand. But Thompson believes that climatic fluctuation also plays a major role in the cattle cycle. He points out that a major part of U.S. cattle production is centered in the Great Plains, where there is a long record of climatic fluctuations that are at least roughly cyclical.

"When there is a period of several successive years of cooler and wetter summers," Thompson says, "farmers and ranchers start increasing their herds. This reduces the number of animals being marketed and causes prices to rise. Thus the number of cattle on ranges reaches large numbers before the weather turns unfavorable."

When drought comes to the plains, feed prices start to rise at the same time that a reduction is occurring in the carrying capacity of rangeland, where the calves are born and feed on grass and other forage plants until they are sent to the feed lots (Figure 10). The result, according to Thompson, is a period of cattle liquidation that lasts about as long as the unfavorably hot and dry summers prevail. But he believes that it should

FIGURE 10 Irrigated pastures like these cushion the impact of climatic variability on cattle raising in the western United States. (United States Department of Agriculture.)

be possible for the livestock industry to adopt a strategy that recognizes the role of climate in the cattle cycle and takes measures to reduce what McDougal calls the "devastating financial results during the liquidation phase." Thompson suggests:

> Assume that the decade of the 80s has a high probability, at least in the part of the United States where most of our cattle feed and forage is produced, of having normal or better weather. Grain prices should be low, and the build-up of cattle will probably begin again. Why not capitalize on what we've learned and build up reserves of feed grain during this period. We should recognize ahead of time that the next decade will have a higher risk of bad weather, and prepare for it.

Although many climatologists would challenge the assertion that climate in U.S. cattle-raising regions is regularly cyclical, Thompson feels that there is enough empirical evidence to provide a basis for the strategy that he proposes:

> Some agricultural economists have perceived this pattern of alternate decades of favorable and unfavorable weather in the Great Plains for at least 20 years. I developed a plan for a group of cattle feeders in 1973 showing that if you had to pick a single year with the greatest drought risk, it would be 1974, and if you chose the three years of greatest risk, they would be '74, '75, and '76. But few of them took advantage of this advice, and several really lost money trying to beat it.

Thompson's main point is that, from the point of view of the Great Plains farmer or cattleraiser, climate is not a completely random variable, even though it can't actually be predicted with any precision. He maintains that, although we can't count on cyclical weather to pinpoint a particular year when there will be low crop yields, we can at least define a 5-year period out of 20 in which the weather risk is greater. Thompson feels that this should be the basis for agricultural strategy for the next 20 years unless the meteorologists learn to do a much better job of seasonal forecasting than they are doing now.

In other words, for a farmer or cattle feeder or anyone else whose livelihood depends on climate and weather, an intelligent guess may provide a considerably better basis for action than the attitude that, because future climatic fluctuations can't be predicted, the best thing to do is to ignore them. In fact, any long-range planning that is done in a climate-dependent situation constitutes what James McQuigg calls "implicit climate forecasting." If farmers or government officials plant a crop or make a farm policy on the assumption that next year's weather will be very much like this year's, they are, by implication, making a climate forecast. McQuigg's point is that anyone who makes such implicit climate forecasts should do it on the basis of the empirical evidence that is available, regardless of how imperfect that evidence may be from a purely scientific point of view. This pragmatic approach is the basis for Thompson's proposed strategy for the cattle industry and is also the rationale for many other strategies that have been proposed for food production, distribution, and reserve policies during the next few decades.

We shall discuss some of these potential strategies in Chapter 8. At this point, there seems to be overwhelming evidence that the climate-crop-consumer machine is a reality, and that climatic change and fluctuation can indeed affect all the people of the world in major ways. In the rich countries, the main impact will be wild fluctuations in food prices when harvests are affected by bad weather, not only within the boundaries of the country itself but in other regions that may be halfway around the world. In poor countries, the effect of climatic anomalies is more likely to be widespread malnutrition and starvation, as the huge North American crop surpluses that have averted major famines in the past can no longer be relied on to cushion the impact of widespread episodes of unfavorable weather such as those we have experienced in the 1970s.

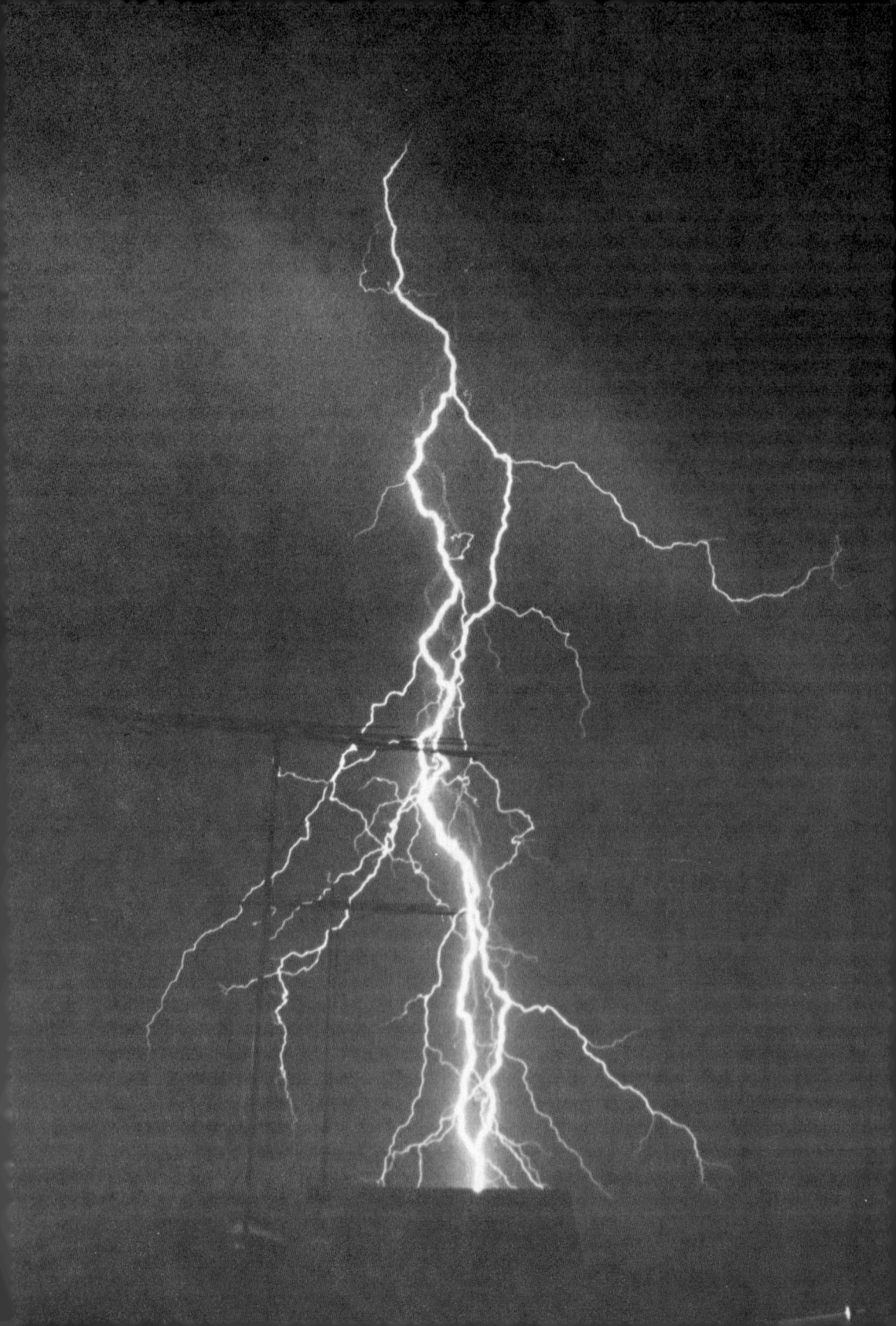

CHAPTER 7

Cloud Seeders and Plant Breeders—
Technological Prospects and Limits

T HE SAN LUIS VALLEY of southern Colorado is a high, arid basin that lies between the San Juan Mountains on the west and the Sangre de Cristos on the east. To the north, the valley funnels down as the two mountain ranges converge at Poncha Pass. To the south, it opens out into the highlands of northern New Mexico. The flat valley floor stands nearly a mile and a half above sea level and covers an area roughly the size of Connecticut.

Several years ago, in March 1973, the principal town of the San Luis Valley, Alamosa, was the setting for a semijudicial proceeding that struck some spectators as a sort of technology-age version of the Scopes monkey trial, in which the arraignment of a schoolteacher in Dayton, Tennessee, precipitated a lengthy and thundering debate on the theory of evolution between William Jennings Bryan and Clarence Darrow. Although the debators at Alamosa were neither as prominent nor as eloquent as those at Dayton, both cases involved large societal questions that went far beyond the immediate issue at hand, and both attracted participants and spectators who came great distances for reasons broader than those that concerned the local people who were arguing over a local controversy.

The occasion for the Alamosa debate was a public hearing held under the provisions of the Colorado Weather Modification Act of 1972, one of the three or four most effective pieces of state legislation that have been passed to regulate or promote weather modification. The Colorado law

Lightning, hail, and hurricanes are among the violent and destructive weather phenomena that atmospheric scientists have tried to tame with weather-modification technology. Cloud seeding has also been used to try to augment beneficial elements of the weather, such as rain and snow in regions where they contribute to water supplies. (Henry Lansford, National Center for Atmospheric Research.)

provides that anyone who undertakes to change the weather within the borders of the state must be licensed by the Colorado Department of Natural Resources and, to conduct a weather modification operation, must obtain a permit each year from that same agency. The decision to grant or deny the permit is made by the executive director of the Department of Natural Resources, with advice from a hearing officer and a committee of five scientists and five farmers and ranchers.

The hearing in Alamosa concerned the application of Atmospherics Incorporated, a Fresno, California, weather-modification firm, to conduct a cloud-seeding program in the San Luis Valley. This was to be a continuation of weather-modification work that had been done in previous years with support from a local organization known as Valley Growers, Incorporated, made up of farmers who grew Moravian barley under contract to the Adolph Coors Company of Golden, Colorado. Coors used the barley to produce the famous beer that is "brewed with pure Rocky Mountain spring water."

The weather has always been a topic of more than passing interest in the San Luis Valley. In spite of the fact that the valley averages less than 7 inches of precipitation a year, many of its people make their living by farming. Beneath the flat alluvial floor of the valley, a thick series of interlayered clays and sands known as the Alamosa Formation slopes down toward the center of the basin. This formation brings water down from the surrounding mountains, feeding artesian wells in the valley. In addition, many farmers have irrigation rights on the Rio Grande and other streams that are fed by melting snow in the high country. Thus although it is a desert by climatic definition, the San Luis Valley is intensely cultivated and produces barley, lettuce, potatoes, and other crops. Ranching is also widespread in the valley; in the early 1970s about 150,000 head of cattle grazed on its arid rangeland and were fed on grain raised in the irrigated fields.

Two weather phenomena are of special interest in the San Luis Valley: rain and hail. If there is even less rain than the normal few inches, the ranchers must use more feed grain and the farmers more irrigation water. Hail is not of much concern to the ranchers, but it can be a big problem for the farmers. In 1971, for example, losses in the valley from hail damage to Moravian barley alone came to over a million dollars.

The farmers who raised Moravian barley for Coors had another problem that was not shared by the growers of feed barley or potatoes. Moravian barley that is to be used for brewing requires what the growers and brewers call a dry harvest. If it gets much rain during the last eight or ten weeks before it is harvested, the barley will not germinate properly during the malting process at the brewery.

Thus if people could manage the weather in the San Luis Valley each summer, there would be at least three conflicting views of what should be done. The ranchers would want all the moisture they could get in any form, even hail, which does little damage to grassland. Most of the farmers would want as much rain as possible, but no hail. And the Moravian barley growers would ask for rain in the spring and early summer, no rain from about the first of July until early September, and no hail at any time. Barbara Farhar, a sociologist who has studied public attitudes toward weather modification, describes such a situation as a "heterogeneity of weather needs."

Until fairly recently, the growers and ranchers of the San Luis Valley, like farmers and herders in many lands over many centuries, accepted hail and drought as inevitable hazards sent by an angry God or an impersonal providence. But in the early 1960s, some of them began to consider that people might be able to take a hand in the workings of the weather by applying a new technology known as cloud seeding.

Cloud seeding was discovered in 1946 by Vincent Schaefer of the General Electric Research Laboratories. Schaefer, who was an assistant to the distinguished physicist Irving Langmuir, made the discovery by a process that he likes to call serendipity—the fortuitous discovery of one thing while looking for another. He had been working in his laboratory with a home freezer he was using as a cold chamber to test different substances that he thought might trigger the formation of ice crystals in moist air. None of them worked, but when he put a piece of dry ice—solid carbon dioxide—into the freezer to help cool it down, a myriad of ice crystals suddenly formed. He had discovered cloud seeding with dry ice. Subsequent tests that Schaefer and Langmuir conducted in the atmosphere proved that dry-ice seeding could stimulate the formation of ice crystals in real clouds.

The seeding worked on clouds made up of supercooled water droplets that were still liquid even though they were colder than 0 degrees Celsius (32 degrees Fahrenheit), the temperature at which liquid water usually freezes. The cloud droplets have a tendency to resist freezing until they are brought to a much lower temperature, as they were by the dry ice, which had a temperature of about $-78°C$ ($-5°F$). Further experiments showed that $-39°C$ ($14°F$) was the critical temperature at which supercooled cloud droplets would become ice crystals. Not long after Schaefer's experiments, Bernard Vonnegut, also of General Electric, proved that crystals of silver iodide would trigger freezing of supercooled cloud droplets, even at temperatures warmer than $-39°C$. The similarity between the crystal structure of ice and that of silver iodide apparently accounted for this phenomenon.

In their cloud-seeding experiments, Schaefer, Langmuir, and Vonnegut were working with a process called ice nucleation that is responsible for the formation of precipitation in many different kinds of clouds in temperate regions such as the interior United States. Winter snow clouds are obviously colder than the nominal freezing temperature of water. But even on the hottest summer day, the upper parts of cumulonimbus clouds—thunderheads—in temperate regions are also considerably colder than 0°C. In such a thunderstorm, tiny particles of certain substances that are naturally present in the atmosphere serve as nuclei for the formation of ice crystals from supercooled droplets. Once a crystal forms, it grows rapidly, taking up moisture from the droplets around it. In a winter cloud layer, the crystal soon grows large enough to fall, and if the day is cold enough, it will reach the ground as a snowflake.

A summer thunderhead often has strong updrafts that hold the ice crystals up until they grow fairly big. When its weight overcomes the force of the updraft, an ice particle falls, and if the day is hot it may melt on the way down, reaching the ground as a big raindrop. But if the updraft is very powerful, and some other conditions are just right, layers of ice can continue to grow on the particle until it becomes a hailstone. Big thunderstorms over the Great Plains often produce hailstones as large as walnuts or even baseballs; the verified U.S. record is held by a jagged 766-gram (1.67-pound) chunk of ice that fell from a Kansas thunderstorm in 1970.

Attempts to modify the weather by seeding clouds are based on the hypothesis that clouds sometimes are deficient in natural freezing nuclei. Such a deficiency may have various effects. Most of the cloud droplets may remain in the liquid state, and the cloud will drift across the countryside without dropping any precipitation. Or, given a certain combination of liquid water content, freezing level, and updraft velocity, much of the water may freeze on a comparatively few nuclei, and each ice particle, held up in the cloud by the updraft, may grow until it finally falls to the ground as a sizable hailstone.

If extra nuclei are injected into such a cloud, by sprinkling it with dry-ice particles or sending a plume of vapor loaded with tiny silver iodide crystals up into it, there may be several possible effects. The extra nuclei may trigger the growth of a large number of particles big enough to fall to the ground as rain. If the quantity of extra nuclei is just right, this might suppress hail as well as stimulating rain. But according to one hypothesis, overseeding—the addition of too many extra nuclei—could trigger widespread glaciation—the sudden freezing of water droplets in the cloud. This would produce a vast number of ice particles too small to fall to the ground and would, in effect, suppress rain.

This last possibility was at the heart of the 1973 controversy in the San Luis Valley. In the late 1960s, the barley growers had engaged a commercial weather modifier to seed clouds over the valley in the summer. This particular operator made extravagant claims for his success in controlling the weather. He said, in effect, that he could turn rain and hail on and off to order, and he claimed that he was providing rain when the barley growers wanted it, suppressing rain during the period when they needed a dry harvest, and suppressing hail throughout the growing season.

By 1972, this operator had been replaced by Atmospherics Incorporated, a firm that had a reputation for skill and forthrightness about what it could and could not accomplish with cloud seeding. But suppression of rain along with hail over the barley fields still appeared to be a goal of the project, and the ranchers were bitterly opposed to it on the ground that it would also affect their grazing land.

The administrative apparatus for carrying out the 1972 weather modification law, with its provision for permits and public hearings, did not become fully effective until late in the summer of 1972. When a hearing was held on July 31, a crowd of 600 supporters and opponents turned out, and the hearing went on well past midnight. The 1972 permit was granted, with the stipulation that no deliberate efforts to suppress rain should be undertaken. But the opponents of the project still believed that rain was being suppressed, intentionally or otherwise. Two weeks after the permit was granted, a radar trailer used in the project was dynamited, with about $50,000 damage. The dynamiters were never apprehended.

In November 1972, during the regular general election, a straw vote on the weather modification issue was taken in five counties in the San Luis Valley. The vote was three to one against weather modification, but it was not legally binding. However, it provoked W. K. Coors, president of the Adolph Coors Company, into writing a letter a few weeks later to the Moravian barley growers urging them to support the weather modification program. The letter began like this:

Dear Grower,
    It has been brought to our attention that those well-meaning people who prompted the straw vote on the "weather management" issue in the recent elections described the current weather management efforts in the San Luis Valley as "being imposed upon the barley grower by Coors." What these people apparently did not tell the residents of the Valley is that without an effective weather management program, Coors cannot and will not continue to buy Moravian Brewing Barley in the Valley.

The letter went on to say that, unless Coors had "good assurance of an adequate weather management program for the 1973 barley growing season," Moravian barley contracts in the valley would be cut back by 20 percent. In each successive year without weather modification, the barley purchases from the valley would be cut another 20 percent, until the amount of barley purchased by Coors in the San Luis Valley was reduced to 10 percent of the company's total needs rather than the 60 percent that it represented in 1972.

Coors' decision that the weather modification project was a necessary ingredient of his company's operations in the San Luis Valley probably stirred up some additional opposition from independent-minded residents of the valley. Some other motivations had gotten mixed into the controversy. For example, at one point some Mormons maintained that the weather modification program represented opposition to God's will.

When the hearing on the 1973 weather-modification permit application convened on March 5, in an auditorium at Adams State College in Alamosa, the lineup of adversaries on the stage appeared to be somewhat unequal. On the Atmospherics Incorporated side was a team of five lawyers—two local attorneys and three imported from Fresno and San Francisco. On the other side, representing an anti–weather-modification coalition that called itself Citizens for Preservation of Natural Resources, sat a single Alamosa attorney, Carlos Lucero. The lineup of expert witnesses also seemed somewhat overbalanced on the side of the weather modifiers; those who testified in favor of granting the permit included William Finnegan, a Navy scientist from China Lake, California; Richard Schleusener, the director of the Institute of Atmospheric Science of the South Dakota School of Mines; Merlin Williams, the director of the South Dakota Division of Weather Modification; and Hugh Winn, an agricultural economist from Colorado State University. However, as it turned out, Lucero's skillful cross-examination of several of these witnesses brought out a good deal of information that weighed against the weather modifiers.

Interested spectators at the hearing included an aide to one of Colorado's U.S. senators, a sociologist from the University of Colorado, two representatives of an East Coast anti–weather-modification group known as the Tri-State Natural Weather Association, and one of the authors of this book (Lansford) who, after the hearing was over, summarized the key questions raised by the hearing like this:

- As long as there is considerable disagreement among experts about the basic feasibility of a particular weather-modification technology, as well as about its possible side effects, should it be applied in large-scale operational projects? The expert witnesses from South Dakota, Schleusener and Williams, answered this question with a definite yes—as long as the technology offers reasonably good prospects for benefits, it should be used. . . .

- If operational weather-modification projects are conducted, should it be only with the explicit consent of the people who live in the area that will be affected? Is it possible to define this area accurately and to specify the effects realistically enough so that the people can make an intelligent decision? The opponents of the permit cited the results of the straw vote held in the Valley last November, which went heavily against weather modification. . . .

- Is it possible to specify the economic benefits that will result from either doing or not doing a particular weather-modification project? This subject was debated at length, and the only conclusion was that there is no easy answer. One of the expert witnesses, an agricultural economist from Colorado State University, said that it would cost around $40,000 to $50,000 just to do a good study of the economic impacts of the San Luis Valley Weather Modification Project. (He was assuming that the physical effects of the cloud seeding can be specified with reasonable precision, which is not a safe assumption at all.)

Both the hearing officer and the weather modification advisory committee recommended that the permit not be granted, and on April 1 it was officially denied by the executive director of the Colorado Department of Natural Resources.

One of the witnesses who testified against granting the permit was Charles Moore, an atmospheric scientist from the New Mexico School of Mines and Technology and a former associate of Schaefer and Vonnegut. He expressed the scientific opinion that cloud seeding with silver iodide can have very widespread effects, and that, in the arid Southwest, one of these may be glaciation in the clouds and thus a reduction in rainfall over large areas.

The possibility that the effects of cloud seeding may extend hundreds or even thousands of miles downwind has been at the heart of many weather-modification controversies. The idea dates back to the late 1940s when Langmuir, who was a Nobel laureate and a highly respected member of the scientific community, put his reputation behind claims for the efficacy of cloud seeding that many scientists found recklessly extravagant and unsupported by convincing scientific evidence.

After Langmuir, Schaefer, and Vonnegut had conducted a few experiments involving cloud seeding with aircraft in the vicinity of the General Electric Research Laboratories in Schenectady, New York, GE's legal department began to get nervous about the possibility of lawsuits. If the cloud seeders did in fact change the weather, the lawyers reasoned, someone was likely to be damaged by the change and might decide to go to court. The outcome was Project Cirrus, a series of weather-modification experiments that used military aircraft and crews, rather than General Electric ones, to do the actual cloud seeding, although the project was still directed by the GE scientists.

In 1948, Langmuir, Schaefer, and Vonnegut moved Project Cirrus from New York to New Mexico, where they began cooperative field work with the New Mexico School of Mines. On October 14, an airplane dropped 2 ounces of silver iodide in burning charcoal pellets from 12,000 feet, and three cloud-seeding runs were made with dry ice. Langmuir claimed that the seeding produced about 0.35 inch of rain over 4000 square miles—a total of about 200 billion gallons of water.

On July 21, 1949, Project Cirrus used ground generators to seed with silver iodide. The generators burned a silver iodide solution, producing smoke that contained billions of tiny silver iodide crystals that drifted up into the atmosphere. Langmuir fired up his generator at 5:30 AM. Three hours later, a cloud began to form about 25 miles south of the generator site. By midmorning, the cloud had become a thunderstorm, and heavy rain was falling. More clouds formed, and more than an inch of rain fell in the vicinity. The generator operated for 13 hours, and Langmuir claimed that he had produced several billion gallons of rain. After calculating the statistical probability that rain might have occurred naturally on those two days, Langmuir announced that the odds were 10 million to one that the seeding had caused the rain, a claim that was received with incredulity by many meteorologists.

Langmuir then began seeding on a regular schedule, on the theory that the periodicity of the weather should correspond to the periodicity of his seeding if the seeding was actually affecting the weather. Langmuir maintained that rainfall as far east as Buffalo, New York, and Philadelphia was being affected by his single silver iodide generator in New Mexico. When he seeded on the first three days of the week, those cities and much of the United States east of New Mexico reportedly had three to ten times as much rain on the weekends as during the week. When he shifted his seeding days, the rainfall patterns appeared to shift with them. This was proof enough for Langmuir that people now had the power to control large-scale weather systems.

But many scientists felt that Langmuir's evidence was less than convincing and that his claims were based more on enthusiasm than on statistically significant data. Weather is notoriously erratic, they said, and natural variability could easily account for the apparent correlation between Langmuir's seeding and the rainfall to the east. The U.S. Weather Bureau and the Air Force conducted their own cloud-seeding experiments in 1948 and 1949, and they found no solid evidence that the seeding was modifying the weather in any consistent way.

Although the meteorological community was generally skeptical of cloud seeding, by the early 1950s, as drought afflicted much of the United States, about 10 percent of the land area of the nation came under commercial cloud-seeding operations intended to increase rain or snow or suppress hail. Farmers, utility companies, and other groups spent several million dollars a year on weather modification during this period. But as the climate grew more favorable for agriculture in the late 1950s, support for weather-modification research and operations lagged.

One response to the impacts on agriculture of the climatic anomalies of the early 1970s was a rebirth of interest in weather-modification technology. Naturally enough, some of the renewed enthusiasm came from scientists and commercial operators who saw new social and economic justification for their cloud-seeding research and operations. This focus on the possible benefits of weather modification for food production quickly became evident on the scientific conference circuit. In January 1975, the newly formed North American Interstate Weather Modification Council sponsored a conference in Denver with the optimistic title: "Weather Modification—A Usable Technology—Its Potential Impact on the World Food Crisis." A few months later, with support from the National Science Foundation, about 50 atmospheric and agricultural scientists met at Colorado State University to try to define "The Present and Potential Role of Weather Modification in Agricultural Production" as part of a National Academy of Sciences study of climatic fluctuation and U.S. agricultural production.

In August 1976, the World Meteorological Organization (WMO) held its second international scientific conference on weather modification in Boulder, Colorado (the first was held in Tashkent, USSR, in 1973). The scientists who presented papers on rain and snow enhancement, hail suppression, and other types of weather modification came from Argentina, Australia, Canada, France, India, Italy, Iran, Israel, Kenya, South Africa, the Soviet Union, Switzerland, and the United States. Weather modification obviously was a subject of more than parochial interest.

The new enthusiasm for weather modification was not limited to scientists and technologists; many political decision makers were also interested in trying to compel the weather to behave in more agreeable ways. In February 1976, Congressman Glenn English of Oklahoma introduced a bill in the U.S. House of Representatives that would have authorized the federal government to pay 60 percent of the cost of cloud-seeding programs established by soil conservation districts in drought-stricken states. This bill died in committee, but later in the year Congress passed the National Weather Modification Policy Act of 1976, which directed the Secretary of Commerce to develop a comprehensive and coordinated national policy on weather modification and a national program of weather-modification research and development.

During the Winter of '77, as drought grew increasingly severe in the western United States, several states established cloud-seeding programs. In January, the snowpack in the Colorado Rockies, which supplies a major part of the water supply for several western states as well as supporting Colorado's ski industry, was more than 50 percent below normal (Figure 1). Governor Richard Lamm assembled a "drought council" of scientists and other experts from the state's universities, research centers, and state agencies and asked them for advice on the most effective ways in which the state might respond to the shortage of precipitation. Professor Lewis Grant of Colorado State University (CSU) recommended a cloud-seeding program using techniques that he and his colleagues had developed and tested in field research in the Colorado Rockies over the past decade. Grant maintained that, given the right conditions, increases in precipitation from orographic (mountain-influenced) snowstorms of 10 to 20 percent could be obtained by seeding, and that any increase of 15 percent or more would almost certainly be cost effective.

Although not all of the members of the drought council favored cloud seeding, Governor Lamm asked the Colorado General Assembly to appropriate $189,200 to support an emergency cloud-seeding program. The bill whizzed through both houses in near-record time—it became law within a week. At the conclusion of the project in May, Grant and John LeCompte, also of CSU, produced a report that concluded that:

> A detailed evaluation of a short-term application program is not feasible since the natural variability of precipitation is larger than the changes in precipitation to be expected from seeding. . . . If the seeding effects were the same during this drought emergency program as those found during similar seeding opportunities during previous long-term orographic cloud seeding experiments, an increase in precipitation of 13% to 19% would be expected.

In other words, although it was not possible to verify with certainty just how much, if any, the cloud seeding had augmented the snowfall, the atmospheric scientists were optimistic about its probable effectiveness. This was good enough for the state legislature, which included provisions for more cloud seeding the following winter in new drought legislation passed in the spring of 1977.

Cloud seeding was also being tried as a remedy for drought in other parts of the world. In 1973, for example, as the Sahelian drought reached

FIGURE 1 Cloud seeding has been used, apparently with some modest success, to increase the snowpack on mountain slopes like these in the Front Range of the Colorado Rockies. (National Center for Atmospheric Research.)

a peak of severity in West Africa, the government of Niger, in the heart of the Sahel, decided to try weather modification. A U.S. firm from Norman, Oklahoma, was engaged to seed clouds from aircraft during September. Later that month, the French Government sent a cloud-seeding aircraft and crew to Niger to augment the effort. Although a little rain fell at several locations during the period, it was impossible, as in many weather-modification operations, to tell how much of it, if any, could be attributed to the seeding. But at a press conference late in September, the government's weather-modification coordinator, Oumarou Youssoufou, said that the results of the cloud-seeding program were "very satisfying."

This sudden and widespread interest in weather-modification technology as a remedy for drought and other weather problems did not come as any great surprise to those who had been observing the weather-modification scene for some time, especially in the chronically water-short Great Plains of the United States. Glenn Lorang, who covers that part of the country for *Farm Journal*, wrote in the February 1972 issue, before the agricultural impacts of that year's climatic aberrations were obvious: "One way or another, a storm is going to break over the Great Plains the next time that area has a serious drought. If it isn't a rainstorm relieving the drought, it will be a storm of demands from farmers and local politicians that Uncle Sam get cracking with cloud seeding."

Also anticipated was controversy over whether or not cloud seeding is a usable technology or one that involves too many scientific uncertainties to be ready for widespread application. In a paper delivered in April 1972 at a symposium on "The High Plains: Problems of Semiarid Environments" held by the American Association for the Advancement of Science, one of the authors of this book (Lansford) said:

> The High Plains may be about to witness a striking example of this disparity between scientific knowledge and public demand. The possibility that this region may be starting into a cyclic drought period similar to those of the Thirties and Fifties has spurred a sharp interest among farmers, public officials, and others in the potential effectiveness of applying weather-modification technology to augment inadequate precipitation. . . . And, in spite of advances that have been made in other areas of weather modification, it appears that we know very little more about the effectiveness of cloud seeding for augmenting rainfall from High Plains convective storms than we knew a couple of decades ago.

What *do* we know about weather modification by cloud seeding? Is it all mumbo-jumbo and reckless optimism, or are there some ways in which it can be used to modify the weather knowledgeably and dependably?

There is one application of cloud seeding about which there is virtually no controversy, over either its efficacy or its use—the dissipation of supercooled fog. When an airport is socked in with fog made up of droplets that are colder than 0°C (32°F), the fog can be dissipated quickly and effectively by seeding with dry ice or silver iodide. However, this is a rather modest application, as about 95 percent of the fogs over airports in the contiguous states of the United States are warmer than 0°C, and there is no fully reliable and practical procedure for dissipating warm fogs over airports.

Second to cold-fog dissipation in terms of apparent success is augmentation of snowfall from orographic clouds—snowstorms that result when moist air moves across a mountain range that lifts and cools it, resulting in precipitation. More than 10 years of experiments in the Colorado Rockies by Lewis Grant and his colleagues at Colorado State University, as well as cloud-seeding experiments and operations in Colorado, California, and other locations, have provided evidence that many atmospheric scientists find convincing that snowfall from orographic clouds can be increased by 10 to 20 percent through careful cloud seeding. In the semiarid and arid regions of the western United States, where a major part of the summer water supply comes from the melting snowpack in the mountains, orographic snow augmentation can be an important application of weather modification technology. But during periods of snow drought, when there is a shortage of snowstorms that produce large snowfalls naturally, it is probable that favorable situations for cloud seeding will also be scarce. Seeding cannot cause snow but can only produce a modest percentage increase in snowfall from clouds that meet a particular set of criteria. The 1977 cloud seeding in the Colorado mountains, for example, began at about the time that the blocking situation that existed earlier started to break down. Thus the results of the seeding probably were more favorable than they would have been earlier in the winter. As Governor Lamm put it: "You can't seed no clouds."

Lewis Grant, who is known internationally for his weather-modification experiments at Climax, Colorado, says that he is often asked whether cloud seeding works. What the questioner usually means, according to Grant, is really: "Will cloud seeding solve my problem?"

"If the problem is drought," Grant says, "then the answer is no."

Although it was Grant's advice to Colorado Governor Richard Lamm, coupled with pleas from farmers, ski area operators, and water conservation groups, that persuaded the governor to ask the legislature to approve the 1977 emergency cloud-seeding program in the Colorado Rockies, Grant did not claim that the cloud seeding would remedy the water shortage. He warned the legislators that cloud seeding is no panacea for drought, and his warning was repeated by state officials, who cautioned that, even with a large number of favorable seeding situations, the program would add only a small increment to the amounts of snow that fell naturally.

When it comes to growing-season rainfall, the most important need of nonirrigated agriculture, such as dryland winter-wheat farming on the high plains, the evidence for the effectiveness of cloud seeding is very inconclusive. Some experiments have shown precipitation increases, some decreases, and some no effect at all. Although some commercial cloud seeders claim consistent success in using cloud seeding to increase precipitation from summer thunderstorms, many atmospheric scientists feel that the evidence does not support these claims.

A number of research projects have been undertaken in recent years to try to learn more about using cloud seeding to augment summer rainfall. Scientists from the NOAA Experimental Meteorology Laboratory have completed more than five years of experimental seeding of Florida cumulus clouds. They appear to have increased rainfall from individual clouds, but they are not sure what is being accomplished over a large area. William Woodley, director of the project, admitted that some scientists warned that the Florida cloud seeders might be redistributing rainfall without increasing it. But analysis of data obtained in 1976 field work indicated that rainfall over the experimental area may have been increased by 20 to 70 percent by the cloud seeding.

Even if the Florida cloud seeders succeed in increasing rainfall, it is uncertain just how applicable their results would be to areas such as the midwestern corn belt and the high-plains wheat regions where drought is a recurring threat to food production. In the mid-1970s, another experiment was begun with the goal of "establishing a verified working technology and operational management framework capable of producing additional rain from cumulus clouds in the semiarid Plains States." This experiment was a cooperative effort between the U.S. Bureau of Reclamation and the states of Colorado, Kansas, Montana, Nebraska, and Texas, with field sites at Colby, Kansas; Miles City, Montana; and Big Spring, Texas. As of mid-1977, this project was not far enough along to have produced any significant results.

Although some encouraging reports on precipitation enhancement programs in several countries were presented at the 1976 international weather modification conference in Boulder, others were very equivocal. For example, Israeli scientists reported rainfall increases of 10 to 20 percent in an experiment that they conducted around Lake Tiberias from 1969 through 1975, and expressed reasonable confidence in the validity of their evaluation. But a report from an operational cloud-seeding program in northeastern India during the summer monsoon seasons of 1973 and 1974, while presenting results that suggested rainfall increases of 16 percent, cautioned that the results of the operation were not statistically significant. And the results of an experiment in central Mexico indicated an unaccountable decrease in rainfall on the days when clouds were seeded by comparison with the amount of rain that fell on unseeded days.

Hail is another weather problem that reduces harvests in many parts of the world (see Figure 2). In the United States alone, hailstorms destroy at least $700 million worth of crops a year, and worldwide losses run to billions of dollars. In the late 1960s, Soviet atmospheric scientists began reporting significant success in using cloud seeding to suppress hail. The United States government decided to support an experiment to investigate the physical feasibility and economic practicality of hail suppression. The National Center for Atmospheric Research was assigned responsibility for managing the National Hail Research Experiment, which included scientists and research equipment from a number of universities and government agencies. Working from a field headquarters near the little town of Grover, Colorado, the researchers studied hailstorms over a large area of Colorado, Nebraska, and Wyoming, at the heart

FIGURE 2　Mr. and Mrs. Clarence Costner hold monster hailstones that fell on their Missouri farm in 1973. (National Center for Atmospheric Research.)

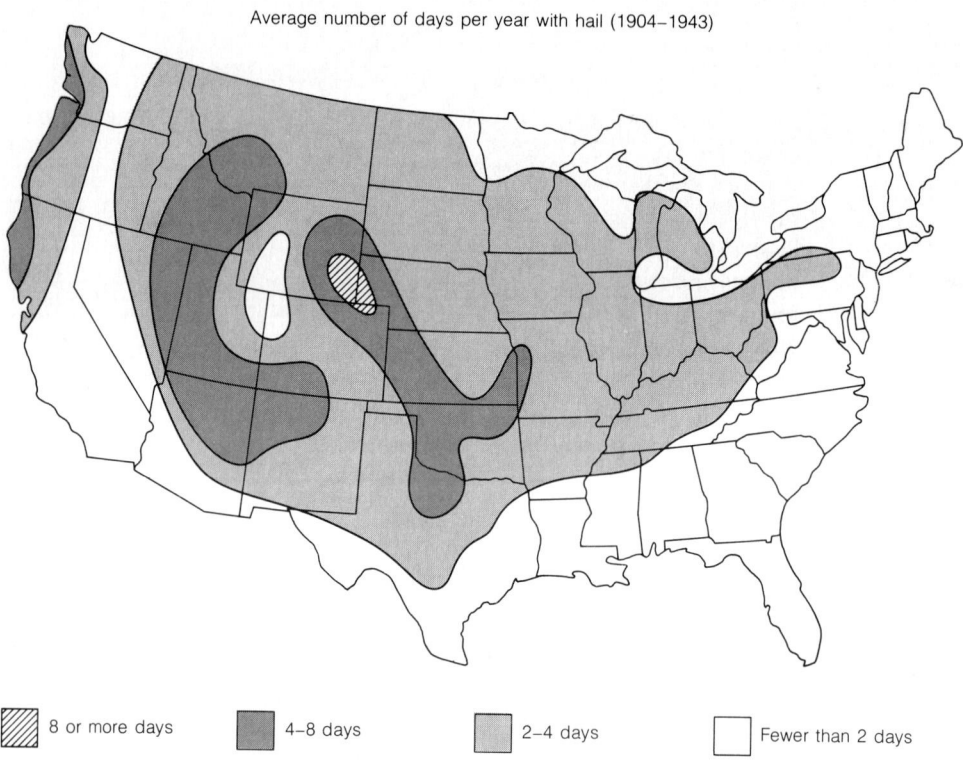

Average number of days per year with hail (1904–1943)

8 or more days   4–8 days   2–4 days   Fewer than 2 days

FIGURE 3   Although hail causes crop damage in many parts of the United States, hailstorms are most frequent and severe in the central part of the country. (National Weather Service.)

of the region known as hail alley because of its numerous hailstorms (Figure 3).

The hail experiment included three years of cloud-seeding tests as well as extensive research on the physics and dynamics of hailstorms, using instrumented aircraft, radar, instruments on the ground, and a variety of other tools. The results were inconclusive as far as hail suppression was concerned, although the experiment produced a great deal of valuable new knowledge of severe thunderstorms. No statistically significant effect of seeding on hailfall or rainfall was detected. This does not mean that hail suppression by cloud seeding is not possible; it simply means that the hypotheses and methods being tested in the National Hail Research Experiment did not prove to be significantly effective in reducing hail.

As the spring of 1977 gave way to summer, the drought in the western United States deepened and appeared to be spreading eastward. In spite of optimistic evaluations of the results of the winter's cloud-seeding program in the Colorado Rockies, it did not appear to many people that weather-modification technology offered much relief from the impacts of drought. Used consistently and carefully as an attempt to maximize the benefits of favorable weather situations, cloud seeding may provide some modest benefits for food production. But it is no panacea for weather and climate problems, and it has the potential of raising a dozen legal, economic, and various other societal problems for every physical problem that it solves. For example, a class-action lawsuit involving millions of dollars in potential claims for loss of life and property was initiated because a 1972 tragic flash flood in Rapid City, South Dakota, occurred not long after cloud seeding had been done in the vicinity. Although the alleged cause-and-effect relationship between the seeding and the flooding appeared tenuous to many atmospheric scientists, the case represented the sort of problem that is likely to develop with increasing frequency if cloud seeding comes into widespread use.

One of the pioneers of numerical weather prediction, John von Neumann, once said that it might prove to be easier to modify the weather than to predict it. Since we clearly have not learned to modify the weather in any consistent and knowledgable way, how much progress are we making in learning to predict the details of weather, or at least the broad outlines of climatic conditions, a season or a year in advance?

In Chapter 5, we discussed two approaches to climate prediction—empirical studies that attempt to determine whether or not certain events can be used as predictors of others without necessarily understanding the whole climate system, and numerical modeling efforts in which the behavior of the system is simulated in the electronic circuits of the computer to try to predict the ways certain elements of the system will act. We indicated that, although modeling may provide the key to climate prediction in the long run, gradual improvement of empirical methods should provide improved seasonal forecasts during the next few years. The two approaches are not mutually exclusive, and recent research efforts have moved in the direction of integrating the theoretical and empirical-statistical approaches.

There is always the possibility that a wild card will turn up in the game of climate prediction. Reid Bryson, the University of Wisconsin climatologist who has been an important, if controversial, figure in the field over the last decade or two, has recently come up with another theory that has caused a good deal of discussion among atmospheric scientists. Although he still regards atmospheric transparency—the effects of airborne dust and carbon dioxide—as the key factor in global temperature changes, Bryson now believes that year-to-year variations in weather patterns may be associated with the earth's wobble as it rotates on its axis. This wobble, known as the Chandler Motion, has a fairly regular cycle of 13 to 15 months, and it appears to produce what Bryson calls a "sloshing of the atmosphere," in such a way as to affect large-scale weather patterns such as those that caused the western drought and eastern cold in the United States during the Winter of '77. To test the theory, Bryson and his colleagues compared it with the real atmosphere. As Bryson described it in an interview that appeared in *The National Observer:*

> We went back and calculated for last year [1976]—very crude calculations— what the circulation in the Northern Hemisphere should have been, month by month, according to our theory. Then we compared it with what actually happened. We did a lot better than chance. Our forecast actually looked quite a bit like what actually happened.

In mid-1977, Bryson's claims for the accuracy of predictions based on the Chandler Motion were fairly modest; he said that they probably offered a 60–40 chance of predicting large-scale weather patterns a year in advance. But he hoped for improvement: "What we're really trying to do is get it up to 75–25, to where there's enough skill so we can make political and agricultural decisions on the basis that we won't have too big a chance of being wrong."

But even a three-to-one chance of making an accurate climate prediction a year in advance leaves a good bit of room for food production to be seriously affected by unexpected climatic developments as well as by local weather events. So in the field of climate prediction, as with weather modification, even the most optimistic projections of present technology offer modest incremental benefits rather than anything like a solution to problems of the impact of weather and climate on food production.

How about approaching the problem from the other direction— breeding crops that will be less vulnerable to the vagaries of weather and climate and producing larger yields and more nutritious food products? Conventional wisdom in recent years has held that these two goals are incompatible. According to this view, if you breed a variety of corn that will mature in a short growing season or that has increased resistance to drought, you must trade off the higher yields that could be attained if you bred for maximum yield with a longer growing season or favorable rainfall. Although there appears to be some truth in this assumption, the situation is not that simple. The Green Revolution grains, which give very high yields under favorable circumstances, appear to continue to outproduce the old varieties even when water supplies, fertilizer, and cultivation practices are less than optimum. But the development of genetic solutions to specific problems of food production is often a difficult and lengthy process that offers no breakthroughs, miracles, or panaceas. The Green Revolution, as we have seen, was more evolutionary than revolutionary. The new high-yield varieties were the end-product of long years of hard work by plant geneticists and their colleagues from other disciplines, not the result of sudden inspiration or discovery. One of the architects of the Green Revolution, Peter Jennings of the Rockefeller Foundation, describes the job of the plant breeder like this:

> A first imperative is the development of varieties that produce abundantly when they are grown in dense plantings and are supplied with fertilizer, water and pesticides. The plants can also be given other desirable properties, such as resistance to insects and tolerance of a wide range of climates, soils and other environmental factors. In principle every heritable characteristic of the plant is subject to the will of the plant breeder.

What this idealized description does not explain, of course, is that, in exercising their will on the heritable characteristics of the plant, breeders must work through generation after generation, often encountering blind alleys and unforeseen complications. It is slow and painstaking work, and new varieties are not created by waving a magic wand. However, new understanding of genetic principles is reducing the element of chance in plant breeding and promises to provide some shortcuts.

Plant breeders work toward other goals besides increased yields and improved resistance to pests and adverse environmental conditions. For example, since cereal grains provide two-thirds of the diet of people in many developing countries, anything that makes those grains more nutritious will help alleviate the world food problem. In his book *By Bread Alone*, Lester Brown points out that the protein in corn, which accounts for a large part of the food consumed in Latin America and sub-Saharan Africa, is deficient in some amino acids that are essential to human metabolism. One of these amino acids is called lysine, and in 1963 plant geneticists at Purdue University discovered a high-lysine corn gene that triggered a number of research efforts aimed at upgrading the protein content of corn and other cereal grains. But as Brown points out, this is not an easy task or one that will be accomplished quickly. As he puts it:

> Analyzing the protein content and quality of currently grown strains is itself an arduous task. Locating genetic lines with desirable qualities involves

painstaking study of thousands of varieties, and breeding and cross-breeding take years. Incorporating high protein content while retaining high yield potential, desirable cooking and eating qualities, and high insect and disease resistance can be incredibly complicated.

In spite of these problems, Brown says researchers hope that eventually the protein content can be increased significantly in varieties of corn, wheat, and other cereal grains suitable for widespread use.

Another possibility is the creation of new grains by cross breeding. Triticale, a cross between wheat and rye, may be the first successful totally new grain species created by the plant breeders. It contains more protein of better nutritional quality than conventional cereal grains, and some varieties of triticale appear to resist both cold and drought better than wheat does. Triticale has come into limited commercial use in the United States, Canada, and some other countries, and work is going on to try to develop other new grains that will be more nutritous and have other desirable features. But in this area, as in nearly every effort in plant genetics, progress will not be rapid, and probably will be measured in decades rather than years. Gradual improvements are the best that can be expected.

There are hazards in overdependence on new plant varieties; one of the most serious is genetic uniformity. Though traditional native varieties of crop plants, called land races by geneticists, are outwardly identical, they include a great deal of genetic variation among the individual plants. For this reason, they are not likely to succumb to outbreaks of plant diseases like the 1970 U.S. corn blight. This gives the land races an advantage for the subsistence farmer, who cannot take the risk of total crop failure in a bad year to gain the benefit of higher yields in the good years; in many countries, if a crop fails completely, the subsistence farmer is not likely to live to raise another one. As Jennings puts it:

> The genetic diversity of the land race can be of great value to the traditional farmer, since it confers at least partial resistance to insect predation and disease, and partial tolerance of environmental stresses such as drought. . . . The net effect of this agricultural system is to give the farmer a measure of security.

Another hazard of the increasing trend toward a few carefully bred varieties of each major food crop is the risk of losing the pool of genetically diverse varieties needed by the plant geneticists to serve as parents for new generations of improved races of plants. Agricultural scientists have recognized this risk and have established banks of diverse varieties of major food crops to maintain sources of seed with varied characteristics for use in future plant-breeding experiments and operations. By drawing on such "gene pools," the plant breeders can respond to food-production problems with species such as the fungus-resistant variety of corn that eliminated the threat of a recurrence of the 1970 corn blight in 1971.

There is little doubt that technology can provide many benefits for maximizing agricultural production in the face of adverse weather and climate. But there is a danger that we will expect too much of the cloud seeders and plant breeders. Faced with widespread outbreaks of unfavorable weather for agriculture, like those that occurred in 1972, neither atmospheric nor agricultural technology is likely to produce any miracle solutions. One, two, or even more consecutive years of bad weather for food production will almost certainly come along sooner or later, and they will have serious adverse impacts on human life that are beyond the ability of technology alone to prevent or to remedy.

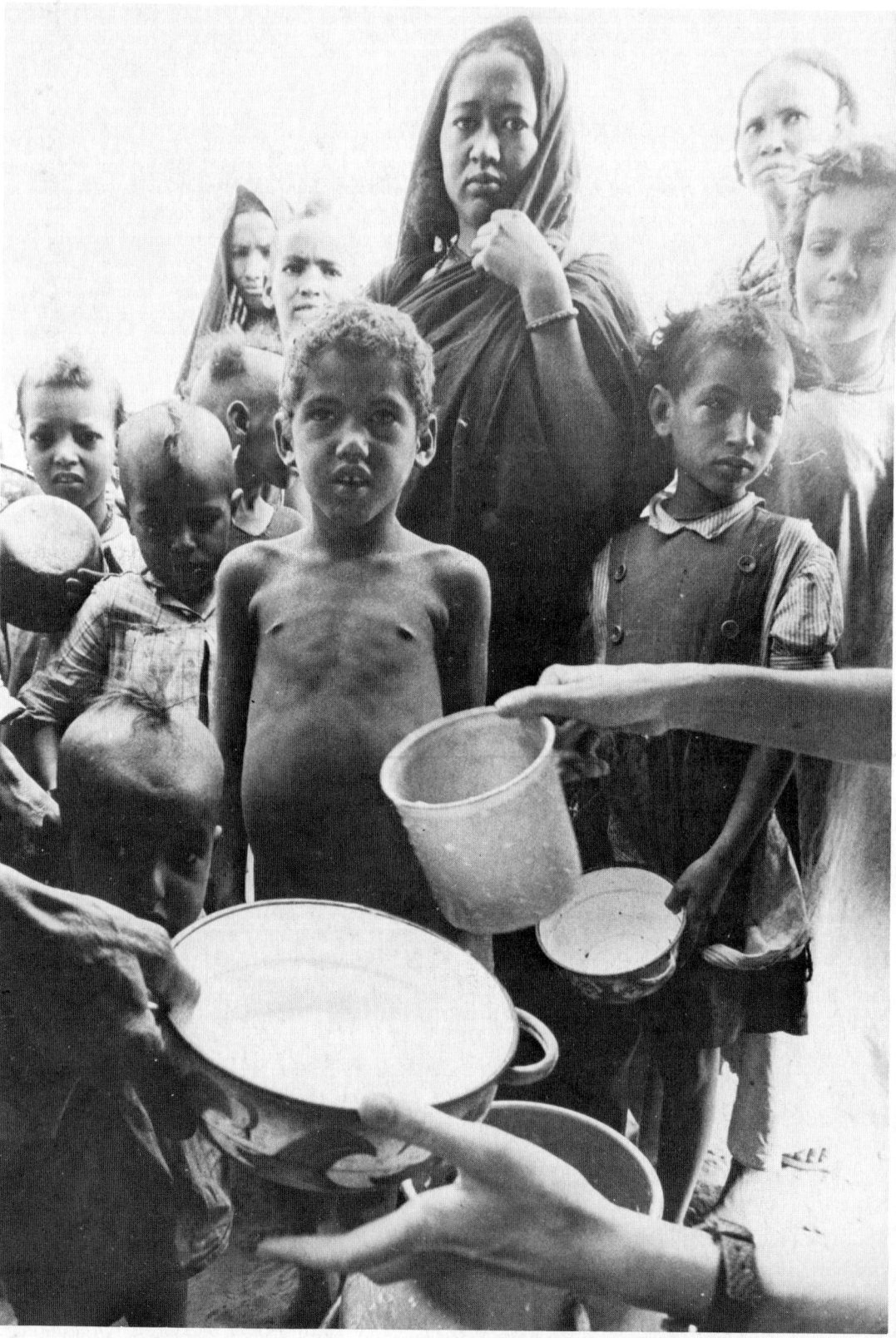

CHAPTER 8

Defining the Problem—Malthusian Pessimists, Social Idealists, and Technological Optimists

WHEN WE CONSIDER the impacts of climatic variation on agricultural production, the key issue does not concern the profits of U.S. farmers or the high cost of food in U.S. supermarkets. The most important question, from a broadly humanistic point of view, is: How can we save the lives of the millions of people who die whenever regional weather problems cause famines and the hundreds of millions that will die if bad growing weather occurs in many parts of the world for several consecutive years? In other words, how can we solve what is usually referred to as the world food problem?

One reason that so many of us have ended up badly confused about the nature and extent of the world food problem is that it is defined so differently by so many people. Some see the hungry people of the world as victims of a single massive problem, overpopulation. These Malthusian pessimists believe the world population is getting close to the ultimate carrying capacity of the planet's natural systems, and some of them advocate strong and coercive measures to force countries with high population growth rates to practice population control. A conflicting view is held by social idealists, who maintain that the problem is neither too many people nor too little food. They believe that we have hungry people because our social, political, and economic systems are not in good working order, and they have a variety of prescriptions for setting things right. Still another viewpoint, technological optimism, defines the fundamental problem as inadequate food production, transportation, and storage. The technological optimists believe that science and technology can be applied to produce sufficient food for many more people than we have now.

Food being distributed in the Lazaret refugee camp, near Niamey, Nigeria, during the Sahelian drought. (United Nations/Gamma.)

Each of these groups is right in a limited way as far as the nature of the problem is concerned. If population growth had not added so many hungry mouths, especially in regions where the ability to produce food is severely limited, we would not have our present world food problem. If we do not find social, economic, and political means to meet the minimum nutritional needs of people everywhere, the human costs will continue to be staggering. If we do not use innovative agricultural technology to increase food production in hungry countries and find the best technological solutions to problems of food storage and transportation over long distances, millions will die from starvation and malnutrition-related disease, achieving population control in a way that seems immoral and unnecessary to most of us.

Solutions to the global food crisis are not likely unless all of these approaches are combined. More effective agricultural and food-handling technology offers long-range benefits for the poor countries, both in providing more food and in achieving economic development that should result ultimately in decreasing rates of population growth. But the benefits are not likely to be realized unless there are also social, political, and economic changes in some of these countries. Professional politicians, in both the developing nations and the advanced industrial ones, play a critical role that has not been given sufficient attention, as Dr. Clifford Wharton, Jr., president of Michigan State University, pointed out in a 1977 article. He believes that the time is at hand for candid international discussion of the political issues of agricultural development. Political leaders in the Third World are beginning to realize that their tenure in office depends to a great extent on their ability to achieve effective agricultural development. The United States is being forced to recognize the political dimension of agricultural development in the developing nations and its relationship to the enormous food-producing capability of North America. As Wharton puts it, "food is power which we must find ways of using wisely, both in our own long-term self interest and in the interest of humanity."

In this chapter, we shall try to analyze the world food problem in a way that synthesizes the views of the Malthusian pessimists, the social idealists, and the technological optimists. Then we shall borrow some lines from each of their scenarios for the future to develop a scenario of our own. We will propose an alternative future that is not as grim as some, if not as cheerful as others, but that takes into account the indissoluble unity of the technical and human elements of the problem.

One of the earliest and most articulate contemporary spokesmen for the Malthusian pessimist point of view was an agronomist and specialist in tropical agriculture named William Paddock. In 1967, with his brother Paul, a retired foreign service officer, Paddock wrote a provocative but pessimistic book entitled *Famine—1975!*, which predicted that population growth and stagnant food production in the underdeveloped nations would converge in 1975 to produce what the authors called a Time of Famines. (This phrase was used as the title when the book was reissued in 1976 with a new chapter explaining why the original timetable had not proved correct, even though the Paddocks' basic thesis remained supportable.)

These widespread famines, the Paddocks maintained, will last for years or decades, and they are inevitable. The Paddocks predicted that, as a result of the catastrophic famines, revolutions, social turmoil, and economic upheavals will occur in Asia, Africa, and Latin America. Although the Paddocks' predictions had been only partially fulfilled by 1978, there is no doubt that they identified a number of problems that still exist and foreshadow potential catastrophic food shortages.

In a chapter entitled " 'Something' Will Turn Up to Avert Famine— Or Will It?" they considered and demolished a whole series of potential "panaceas" that were highly regarded in the 1960s. These included synthetic foods, hydroponics, desalinization, food from the oceans, and several other solutions, some political and economic as well as technological, including "the panacea of the unknown panacea"—the possibility of some unexpected development or discovery that might provide a solution to the problem of converging population and world food supplies. The Paddocks concluded that no panacea, known or unknown, could solve the problem, and that the Time of Famines was inevitable.

In spite of the widespread optimism triggered by the Green Revolution, others shared the Paddocks' dark vision of the future. In his 1968 book *The Population Bomb,* Paul Ehrlich wrote: "The battle to feed all of humanity is over. In the 1970's the world will undergo famines— hundreds of millions of people are going to starve to death in spite of any crash programs embarked on now."

But even the Paddocks and Ehrlich did not predict famine in the United States and other affluent countries. The problem that they foresaw for those nations was moral rather than physical. As the Paddocks put it: "The United States, even if it fully cultivates all its land, even if it opens every spigot of charity, will not have enough wheat and other foodstuffs to keep alive all the starving. Therefore the United States must decide to which countries it will send food, to which countries it will not."

As a guide to making these hard decisions, the Paddocks recommended a method known as *triage.* The term, which comes from the French verb *trier*—to sort—was used by military surgeons during World War I to describe the technique they used when the wounded hopelessly outnumbered the medical people available to care for them. The casualties were sorted into three classes, described by the Paddocks like this:

1. Those so seriously wounded they cannot survive regardless of the treatment given them; call these the "can't-be-saved."

2. Those who can survive without treatment regardless of the pain they may be suffering; call these the "walking wounded."

3. Those who can be saved by immediate medical care.

The Paddocks proposed that in the Time of Famines the United States would be in the position of the battlefield surgeon faced with more wounded soldiers than it was possible to treat. Without enough food to save all of the starving people in all of the hungry countries, the United States, according to the Paddocks, should practice triage. It should send no food at all to two classes of countries—those in which the population growth trend has already passed the agricultural potential and in which inadequate leadership and other divisive factors make catastrophic disasters inevitable, and those which have the necessary agricultural resources to produce enough food or foreign exchange to buy food from abroad. The first group is the "can't-be-saved" countries and the second is the "walking wounded."

The third class of countries, corresponding to the wounded soldiers who can be saved with immediate medical care, are described as "Nations in which the imbalance between food and population is great but the *degree* of the imbalance is manageable . . . in the sense that it can give enough time to allow the local officials to initiate effective birth control practices and to carry forward agricultural research and other forms of development."

In addition to the chances of survival of the nation itself, the Paddocks suggested some other criteria that could be applied; for example, whether the nation's survival will:

1. help maintain the economic viability and relative prosperity of the United States during the Time of Famines;

2. help maintain the economic stability of the world as a whole;

3. help create a "better" world after the troubles of the Time of Famines have ended.

Triage is also the basis of a scenario that Ehrlich developed in *The Population Bomb*. His target date was 1974, when he visualized the United States realizing that the food-population balance in large areas of

Asia, Africa, and South America was so badly out of control that they could not attain self-sufficiency. In this scenario, the United States begins its policy of triage by announcing it will no longer send food to India, Egypt, and some other countries that appear to be in a hopeless condition. Here are some highlights of the subsequent hypothetical sequence of events:

> Famine and food riots sweep Asia. In China, India, and other areas of Asia, central governments weaken and then disappear. . . . Famine and plague sweep the Arab world. . . . Most of the countries of Africa and South America slide backward into famine and local warfare. . . .

In Ehrlich's scenario, the United States, Canada, Russia, Japan, Australia, and the Common Market countries work through the United Nations to set up a program of "area rehabilitation" that will involve simultaneous population control, agricultural development, and limited industrialization, to be carried out in parts of Asia, Africa, and South America. This plan is to be initiated in 1985, when "the major die-back will be over." Ehrlich sees this as the most optimistic of several scenarios for the future, but he says that it calls for "a maturity of outlook and behavior in the United States that seems unlikely to develop in the near future."

In some ways 1968 was a simpler time than 1978, and Ehrlich did not foresee a number of factors that make it highly unlikely that the United States could stand back in "mature" detachment and watch much of the world disintegrate into famine and anarchy. The Arab world would cut off our oil long before it descended into famine and plague for lack of U.S. grain, and terrorists of all descriptions would almost certainly hold cities like New York hostage with real or imaginary nuclear devices or other threats. Even if the United States were able to develop the "maturity of outlook and behavior" required for triage, it is extremely unlikely that we would get away with it.

Triage is usually contemplated as a response to a desperate situation—a general Time of Famines in which the United States must allot a limited amount of food aid to a few out of many hungry countries. Another concept, usually called lifeboat ethics, suggests that we should deliberately discontinue food aid immediately to countries that do not curtail runaway population growth, on the grounds that this is the most effective way to avoid much greater catastrophes in future years.

One of the principal proponents of lifeboat ethics is Garrett Hardin, Professor of human ecology at the University of California at Santa Barbara. Hardin bases his position on what he called "the Tragedy of the Commons" in an article that appeared in *Science* in 1968. In a more recent article published in 1974, he described this concept thus:

> Under a system of private property the man (or group of men) who own property recognize their responsibility to care for it, for if they don't they will eventually suffer. A farmer, for instance, if he is intelligent, will allow no more cattle in a pasture than its carrying capacity justifies. If he overloads the pasture, weeds take over, erosion sets in, and the owner loses in the long run.
>
> But if a pasture is run as a commons open to all, the right of each to use it is not matched by an operational responsibility to take care of it. It is no use asking independent herdsmen in a commons to act responsibly, for they dare not. The considerate herdsman who refrains from overloading the commons suffers more than a selfish' one who says his needs are greater. (As Leo Durocher says, "Nice guys finish last.") Christian-Marxian idealism is counterproductive.

Hardin sees the world's food-production systems—its fisheries, pastures, and farmland—as a commons that is headed for destruction if the poor nations with rapidly growing populations are given continued access to it through food aid with no accompanying demands for population restrictions. He proposes compulsory birth control for all because, as long as it is optional, those nations that do not control population growth will continue to have increasing numbers of people to the detriment of an overpopulated world as a whole, including the countries that have slowed population growth.

Hardin is convinced that the responsible nations—the ones that have reduced population growth—should adopt lifeboat ethics. The affluent industrial nations, he says, are figuratively drifting in lifeboats. The poor nations are swimming in the surrounding ocean, many of them about to drown. With the lifeboats already loaded to capacity, what should their passengers do? Try to save the poor wretches in the water? If they do, Hardin says, "The boat swamps, everyone drowns. Complete justice, complete catastrophe."

Hardin disparages the concept of a world food reserve that has been proposed as a way to cope with periodic food shortages in the poor countries. He maintains that it will only escalate the problem through what he calls the ratchet effect. Without the world food reserve or some other external source of food for emergencies, he says, the population of a nation would go through a cycle in which the number of people rises until a crop failure comes, then drops back to a lower level that can survive on the available food, only to rise slowly again until another crop failure reduces the food supply again. Large inputs of food at the time of the emergency act like the pawl of a ratchet, preventing the population from dropping back in response to the reduced carrying capacity of the nation's domestic food-production system. Thus each emergency will be greater than the last one, Hardin maintains, until finally "the process is brought to an end only by the total collapse of the whole system, producing a catastrophe of scarcely imaginable proportions." Hardin approvingly quotes the third-century theologian Tertullian, who wrote: "The scourges of pestilence, famine, wars, and earthquakes have come to be regarded as a blessing to overcrowded nations, since they serve to prune away the luxuriant growth of the human race." Hardin clearly believes that denial of food aid to any country that does not limit its population to the number of people that can be fed by the nation's own resources, in bad years as well as in good ones, would be a blessing not only to the overcrowded nations but to the rest of the world as well.

But others are not convinced that the dead hand of Malthus is guiding humanity toward an inevitable food catastrophe. Those who hold the viewpoint that we call social idealism define the world food problem in terms of political and economic choices. They believe that efforts to build world food reserves, increase food production, and control population growth are treating the symptoms rather than the fundamental cause of hunger.

In its broadest form, social idealism views the economic gap between the rich and the poor nations as the root cause of world hunger. Many leaders of Third World nations see an inevitable global conflict between the rich and the poor nations. As *Time* magazine described it in 1975:

> On one side are two dozen or so industrialized, non-Communist states whose 750 million citizens consume most of the world's resources, produce most of its manufactured goods and enjoy history's highest standard of living. Demanding an even larger share of that wealth are about 100 underdeveloped poor states with 2 billion people—millions of whom exist in the shadow of death by starvation or disease.

These Third World leaders accuse the affluent nations of wastefulness, arrogance, exploitation of natural resources, imperialism, and neo-colonialism, and they are demanding changes as a matter of human justice. For example, Julius Nyerere, president of Tanzania, says, "I am saying that it is not right that the vast majority of the world's people should be forced into the position of beggars, without dignity. We demand change, and the only question is whether it comes by dialogue or confrontation."

The Third World nations have demanded a new international economic order, which, stated simply, means that the rich nations must make major sacrifices to benefit the poor ones. This position has been supported by numerous resolutions passed in the United Nations General

Assembly, where the Third World nations constitute a majority. The United States and other Western countries have responded with some bitterness at times; Daniel P. Moynihan, former U.S. Ambassador to the United Nations, described the Third World demands as rooted in "the politics of resentment and the economics of envy."

Nevertheless, many Westerners are distressed by the shocking contrast that exists between the lives of most of the people of the rich nations and most of the people of the poor ones. Among the efforts to formulate some specific responses to the problems raised by the militant U.N. resolutions was a study commissioned by a private international organization, the Club of Rome, and headed by Jan Tinbergen, a Nobel laureate in economics. The report that came out of this study, entitled *Reshaping the International Order,* recommended that the rich countries support an international program aimed at the following targets, to be achieved by the year 2000:

> An average life expectancy of 65 years or more, compared with the present 48 years in the poor countries.
>
> A literacy rate of at least 75 percent, compared with the present 33 percent in the poor countries.
>
> An infant mortality rate of 50 or less per thousand births, which would be less than 40 percent of the present average for the poor countries.
>
> A birth rate of 25 or less per thousand, compared with the present 40 per thousand in the poor countries. (Grant, 1977)

Whether or not one agrees with the appropriateness or feasibility of these particular goals, they clearly represent concrete and humanitarian objectives toward which the rich nations could work without feeling like respectable and prosperous citizens set upon by Third World highwaymen. And as far as the world food problem is concerned, it seems highly probable that the kind of development programs that would achieve the goals proposed by Tinbergen and his colleagues would also provide adequate food for the people of the Third World countries.

In the meantime, however, there are many major short-term problems that must be dealt with by individual groups and nations. Should the United States Congress, for example, continue to provide authorization for food-aid programs similar to those established more than 20 years ago by Public Law 480? Should we continue to finance and encourage agricultural development programs like those that brought the Green Revolution?

Some social idealists believe that imported food and technology from the United States have done a great deal more harm than good in terms of helping to feed the people of the poor nations of the world. Imported food, they believe, has removed the incentive for the countries that receive it to develop their own agricultural systems, and imported technology focused on increasing food production, such as that introduced by the Green Revolution, has replaced workable labor-intensive subsistence farming systems with mechanized farming operations that demand large amounts of capital and land and that are highly dependent on imported petroleum, fertilizer, and pesticides.

Frances Moore Lappé and Joseph Collins, co-directors of a group known as the Institute for Food and Development Policy, have expressed this viewpoint provocatively in a number of articles and in a book, *Food First*, that they wrote with Cary Fowler. They say, in effect, that the best thing that the rich countries can do for the poor ones is get out of their way and let them feed themselves. Lappé and Collins are particularly critical of agricultural development programs designed to increase production by applying any technology that will produce higher crop yields. As they see it:

> The production focus quickly becomes synonymous with "modernizing" agriculture—the drive to supply the "progressive" farmer with imported technology: fertilizer, irrigation, pesticides and machinery. The green revolution seeds only reinforce this definition of development because their higher yields depend heavily on these inputs. Agricultural progress is thus transformed into a narrow technical problem instead of the sweeping social task of releasing vast, untapped human resources.

According to Lappé and Collins, this comparatively simple definition of the food problem in technological terms has had great appeal for governments, international lending agencies, and foreign assistance programs because it promises quantifiable results from the expenditure of large but specific sums of money. The result, they say, is that "this influx of public funds quickly turns farming into a place to make money—sometimes big money. To profit, however, one needs some combination of land, money, credit-worthiness and political influence. This alone eliminates most of the farmers throughout the world."

In short, according to Lappé and Collins, the result is that the rich get richer, and many small farmers lose their land and are unable to find jobs on the large farms, where mechanization has reduced the need for human labor. Many small farmers drift into the urban slums, where they remain jobless and hungry. Thus, to quote the title of an article by Lappé and Collins that appeared in the United Nations *Development Forum*, "More Food Means More Hunger."

Lappé and Collins, along with many other social idealists who share their views, believe that there is no developing country in which the food resources could not feed the local people. They maintain that, if the United States and other rich nations would disengage themselves by abandoning food aid and development programs that reinforce the situation in which agriculture is controlled by a small, wealthy elite group in many developing countries, social and political changes can occur that will return food production to the people. Social revolutions rather than green revolutions, they believe, are the only solution to hunger in the poor countries of the world. Adequate food will automatically become available to the people if the social, political, and economic problems are solved.

Other critics disagree sharply with the social idealists, maintaining that their vision of the future is completely reversed, and that increased food production is the key to social and economic improvements in the poor countries. They believe that exporting agricultural knowledge to hungry nations will be much more useful than exporting agricultural products (Figure 1). According to Sterling Wortman, a plant geneticist and vice president of the Rockefeller Foundation:

> An all-out effort to increase food production in the poor, food-deficit countries may be the best means of raising incomes and accumulating capital for economic development, and thus for moving the poor countries through the demographic transition to moderate rates of population growth.

Wortman, who has worked in a number of the Rockefeller Foundation's agricultural programs since 1950, is an articulate spokesman for the point of view that we call technological optimism. Critics of this approach claim that the Green Revolution, which was regarded by many as a technological solution to the world food problem, was far from an unqualified success, particularly in terms of some of its social and economic impacts. They point out that economic inequities between large and small farmers in some countries were increased by the new technology, with its requirements for capital to provide fertilizer, pesticides, and irrigation pumps, and that the emphasis on grain production reduced production of some other crops that are more nutritious. The harshest critics of the Green Revolution charge that it was an ill-advised technological adventure at best and a tragic hoax at worst.

There does not seem to be much basis in fact for such ultranegative views—the Green Revolution clearly provided food for many hungry people, and the fact that government decision makers did not take advantage of the time it allowed to solve some serious social and

FIGURE 1   Klaas Kuiper, a soil fertility expert from the Netherlands, examines samples of fertilized and unfertilized rice in Indonesia. Kuiper was sent to Indonesia by the United Nations Food and Agriculture Organization as part of a demonstration program instituted by the United Nations Development Program. (United Nations/Food and Agriculture Organization/Botts.)

economic problems is not a fault of the program. As Nicholas Wade wrote in *Science* magazine in 1974: "A major impediment to assessing the present state of the Green Revolution is the rhetoric that has accreted around it." Excessively high expectations for the efficacy of the Green Revolution as a solution to the world food problem gave way to disillusionment that has sometimes obscured the very real and substantial benefits that the new seeds and technology brought to many people.

The most important lesson of the Green Revolution seems to be the need to avoid the assumption that new technology alone will solve food problems. To repeat Haas and White's phrase that we quoted in Chapter 2, "It is not a question of more technology or less technology, but of technology in balance." In his introduction to a special issue of *Scientific American* on food and agriculture, containing a dozen articles that generally embodied technological optimism, Wortman cautioned that radically new public policies will be needed in both the rich nations and the poor nations for new technology to produce the benefits that he envisions.

Wortman warns against three technological "nonsolutions" to problems of food and hunger that could divert attention and resources from real solutions. The first nonsolution is larger harvests in countries that are already producing large food surpluses, for instance, the United States, Canada, and Australia. Continuing to supply food from these bread-basket nations to countries that have neglected their own agriculture simply "allows governments to put off the tedious and unglamorous task of helping their own people to help themselves."

A second nonsolution is the introduction of Western-style, large-scale mechanized farming into developing countries. In small countries with limited arable land and large populations, labor-intensive agriculture on a small scale may be a better strategy for food self-reliance than mechanized agriculture, when all factors are considered.

A third technological nonsolution, which was also cited by the Paddocks as one of their nonexistent panaceas, is the production of synthetic foods. Unlike the production of food in the countries where it is needed, which provides income for small farmers and their families, the manufacture of synthetic foods in the industrialized nations would not have any major social benefits; hungry people would not have money to buy it, even if it were cheaper than natural foods.

W. David Hopper, a Canadian scientist who has been deeply involved in international agricultural development programs for the last several decades, agrees with Wortman that the introduction of new farming systems, based on the careful and selective application of technology to the agricultural needs of developing countries, is the key to providing more food and, at the same time, improving the economic health of the hungry nations. According to Hopper:

> The tropical and subtropical resources of the developing countries are now mainly exploited by farming techniques that have remained almost unchanged for centuries. . . . It is now clear, however, that where modern plant varieties and farming techniques are introduced, farmers succeed in wrestling from their land single-crop yields two or three times as large as the traditional return; multiple cropping of two or three crops on the same piece of land—something that is peculiarly feasible in the Tropics and sub-Tropics—gives yields from four to eight times larger than traditional ones.

A good deal of agricultural research is now directed not only toward achieving higher agricultural yields per se but also toward developing technology that small farmers, without major capital resources or ready access to credit, can use effectively. A good example of this approach is seen in the work of the International Center for Tropical Agriculture (CIAT) at Cali, Colombia. One of eight major international agricultural research centers supported by several governments and foundations, CIAT conducts research focused on the agricultural and economic development of the lowland tropics, with special emphasis on the needs of rural and urban low-income groups. According to CIAT director John Nickel:

> If the income of small farmers is increased . . . unit costs of production will be lower and there will be more food available at reasonable prices to the urban population. Similarly, the small farmers will be more likely than more affluent segments of society to spend their additional income on the types of products which can be produced in local labor intensive industries rather than imported goods, thus increasing the opportunities for what the urban poor need most: jobs. Following this chain of events, when the urban poor are employed and their income improves, urban markets for agricultural products are expanded for the small farmer, thus completing the circle (Harrison, 1977).

A major thrust at CIAT is toward breeding plant varieties that are adapted to the environments where they will be grown rather than developing expensive technologies to resist environmental factors. As one CIAT scientist, Bert Grof, puts it: "If you alter the environment enough you can grow geraniums at the north pole. But we try to adapt plants to their environment, breeding strains that will grow in poor soil rather than applying bags and bags of fertilizer."

In other words, CIAT tries to breed crops that use less fertilizer because they are more efficient in utilizing the elements that are available in the soil, need less pesticide because they resist insects and diseases more vigorously, and require less herbicide because they are cultivated in ways that hold down the growth of weeds. Instead of grains, which generally are suited to large-scale farming, CIAT has concentrated on crops that are suitable for small farms in the tropical lowlands: beans, cassava, and legumes for grazing cattle. The center has made considerable progress in increasing yields of these basic tropical crops without requiring increased capital investments for the small farmer.

But regardless of how appropriate new seeds and techniques may be in terms of the climate, soil, and other physical factors, they cannot simply be plugged in and started up. They need to be supported by what the development specialists call infrastructure—good roads, vigorous agricultural research institutions and extension services to transfer new knowledge from the researchers to the farmers, and the economic resources and government policies that are needed to support a strong agricultural establishment (Figure 2). Most of Africa, Hopper points out, is still in the earliest stages of traditional farming, with very low productivity. However, a few areas in Africa that were developed by European farmers are highly productive, not because of ethnic superiority or some intrinsic efficiency of colonialism but because the Europeans built their own infrastructure. The high productivity of the European-run agricultural operations in Africa, Hopper maintains, "can be traced to the large investments made in roads, supply depots, markets, farm equipment, research and extension and farmers organizations, and in the economic institutions and government policies necessary to make all those elements function."

Thus technological optimists like Wortman and Hopper see two aspects of agricultural development that must go hand in hand to achieve the dramatic increases in food production they believe are possible in many of the poor and hungry countries. First, technologies must be developed that are specifically suited, economically and socially as well as ecologically, to the needs and conditions of the countries where they are to be applied. Second, the governments of those counries must make solid and continuing commitments to agricultural development and self-reliance; there is no way that the necessary infrastructure can be built without the support of the political and economic power structure of the developing nation itself. This view implies a strong and continuing commitment by the developed countries to support agricultural and economic development in the hungry nations rather than keeping them dependent on imported food. It also calls for agricultural development programs based on appropriate technology, such as that being developed

FIGURE 2   Agricultural extension stations like this one at Banju Biru in Central Java are an important part of the infrastructure that must be established if developing countries are to make significant advances in food production. (United Nations/Food and Agriculture Organization.)

by CIAT, rather than a production-at-all-costs approach that may make the developing country almost as dependent on outside resources as it would be if it continued to rely on imported food.

Each of these views of the world food problem—Malthusian pessimism, social idealism, and technological optimism—contains elements of truth. Overpopulation; inequitable social, political, and economic arrangements; and low agricultural productvity all clearly contribute to hunger. The danger lies in the conviction that a single aspect of the problem is so overriding that attempts to deal with any others are simply a waste of time and money.

The premise that many human lives can be saved and much human misery avoided by a diversity of efforts is the basis for a position that, for want of a better term, we call pragmatic humanism. This view does not reject technology because of the real or supposed shortcomings of the Green Revolution, nor does it say that any means is justified to achieve such goals as reducing world population growth. It simply holds that the sum of human knowledge provides us with a great number and variety of tools for fighting hunger and that different tools will be useful for doing different jobs in resolving the problem as a whole.

The whole-system approach that we call pragmatic humanism appears to us to be the most workable approach to the world food problem, especially in view of the great uncertainties imposed by the climate mandate. In the remaining pages of this chapter, we shall discuss some of the whole-system alternatives to the population-food-climate problem that appear to combine both the greatest potential benefits for improving the human condition and the highest probability for success. We shall reject two classes of approaches: those that propose large involuntary sacrifices of human life for some theoretical higher good and those that, regardless of their abstract merits, appear to have an extremely low probability of successful execution in the real world.

From the agricultural technology standpoint alone, there appears to be no insurmountable problem in increasing world food production to several times its present level, although the ecological implications have not been examined in any comprehensive way. Peter Buringh has reported such projections based on models developed by a highly respected research group at the Agricultural University of Waginingen in the Netherlands. These models show that, from the technical standpoint, world agricultural productivity could be increased tenfold even with a substantial reduction in the area of the world devoted to agriculture and an increase in forested areas. Why, then, has agricultural productivity lagged so far behind much more modest targets set by the United Nations Food and Agriculture Organization (FAO)? Only 21 out of 62 countries reached the 3.4 percent annual food production increase set by FAO as a goal for 1961–1975; in the same period, population growth exceeded food increases in 42 out of 92 developing countries.

The answers to this question are not easy to come by. They involve political issues, education, and technology, as well as finance, social justice, and the contest of rival economic systems. No technological fix will solve the problem of food production that lags behind population growth, but without a sound technological base, no solution will emerge.

Our view of the world food situation is strongly shaped by our conviction that the following statements are fundamentally correct:

First, the world's best agricultural lands, by and large, are already under cultivation, although it may be possible to double the total arable land area. The cereal grain crops that comprise 75 percent of the world's food are grown on 66 percent of the present agricultural land. However, productivity of some of these best lands now under cultivation could be greatly improved, probably at a significantly lower cost than opening new lands. Total food production on the world's best agricultural lands can almost certainly be quadrupled by application of advanced agricultural practices.

Second, fertile and reliably productive agricultural acreage is a nonrenewable resource, and substantial areas of the best agricultural or forest lands are lost every year to urban development, road and airport building, salinization, wind and water erosion, deforestation, and other preventable factors. Resolute action can minimize these losses even while food and fiber production increases.

Third, marginal lands are being pressed into agricultural production, in some areas at a serious cost to forests and grazing land. Timber and firewood are becoming desperately scarce in some countries where half of the wood production is used for fuel and is an essential energy supply of many poor and hungry nations. The growth of deserts is accelerated by extension of agriculture and grazing into already marginal zones, especially in arid regions and on steep, rocky slopes. Great improvements in grazing land and forest productivity are possible on marginal lands with modern technology, including techniques for controlling the spread of deserts.

Fourth, modern agriculture, with machines, irrigation, and heavy use of fertilizer, represents a very small fraction of total energy consumption, even in countries with highly mechanized and energy-intensive agriculture. Even in an increasing energy pinch, agriculture should be provided with the energy that it needs because of the primary human importance of food.

Fifth, expansion of modern agricultural practices to increase food production will not greatly increase pesticide and herbicide pollution. Nearly half the pesticides used in the United States go into parks, golf courses, and lawns, and almost half of the insecticides used on crop lands go to cotton and not to food crops.

In the light of these and other considerations, the pragmatic humanist is confronted with a profound challenge. On one hand, it appears that the world food crisis is *not* one of insufficient total food-producing capability. Even a doubled population will not begin to strain at the resource limits. The Malthusian pessimists are guilty of oversimplification. The earth as a whole is *not* straining at ultimate limits of food-producing resources; yet the sheer numbers of hungry people increase, proving that the very real prospects of technological optimism are insufficient. Are the social idealists right, after all, in believing that the problems will be solved only by reform of political and economic systems? In our view, the scenario that will resolve the world food crisis involves all three components; of these, perhaps the political and economic ones have been most neglected by those seeking solutions to the food crisis.

There is no reasonable doubt that world population must ultimately reach a zero-growth level. The planet can sustain fertility and the ability to recover from pollution only for some finite upper limit of population. What is this number? The answer is not clear. Some responsible groups believe that the optimum world population has already been exceeded, while others argue that world population can expand fivefold before we are in trouble. However, there is increasing consensus that whatever the number may be, there is a fundamental human right for all of the world's people to have access to basic human needs for a healthy and secure existence, regardless of race, political system, or geographic location. Responsible leaders in every nation are recognizing the need for a redistribution of essential resources to meet minimum human needs— and of these food is high on the list. It is likely that the poor nations with least access to the world's present material goods will be most militant in their defense of their rights to increase their numbers as well as their access to the world's goods and resources.

Our view is that world population may well climb to 8 to 10 billion before a no-growth level is reached and that it is essential to plan ways to provide this number of people with food and other basic needs such as health care, education, and gainful employment. We are convinced that technology can make this goal possible if there is the social and economic will, but we do not underestimate the political adjustments this will necessitate.

In our own nation, for example, politicians will be hard put to stay in office if they espouse such unpopular policies as tough energy constraints that cause higher energy prices, strong price supports for increased agricultural production that result in higher consumer prices for food, heavy taxation for food and development aid to poor nations, and a host of other measures that may be necessary if a revolution in global equity is to come about. With the short cycles of political tenure, practical politicians find it difficult to espouse the kinds of long-range social and economic strategies in our foreign policy that will lead to self-reliance for the developing nations. Crisis aid in a catastrophe is politically easier than a long-range plan to avert catastrophe. Somehow this pattern must change.

In the developing nations themselves, the political constraints are often even more discouraging. It is hard for politicians in these countries to espouse policies to reform land ownership and management when such measures threaten the status of influential elites. Yet such reform is essential to self-reliance in many regions. Measures to use tax revenues to improve the amenities of rural life in order to increase agricultural efforts are difficult to implement when they are at the expense of comparable commitments to urban centers. Consequently, there is a lag in essential infrastructure developments such as roads, agricultural financial credit at reasonable rates, transport facilities, storage and processing plants, communication networks, educational and agricultural extension services, and the like. Without these, food self-reliance for the developing world is an illusory hope. The risk of food imperialism by food-rich nations such as the United States and Canada can be a serious deterrent to cooperation, especially if cooperation involves outside insistence on population control for the developing nations—an issue with explosive racial overtones.

What strategies are possible? The picture is not all bleak. First, leaders in many developing countries are recognizing the importance of agricultural development as a key to their political futures. Second, the United States is demonstrating growing recognition of the world responsibilities that go with the vast North American food-production capability, and we are increasingly aware of the need to use our food power in ways that are in our own intelligent self-interest and are also concerned with overall human welfare. Indeed, many of our leaders are convinced that our entire social-economic system is unlikely to survive if one-fourth of the world's people continue to live under the stress of chronic malnutrition.

Thus we are convinced that the following scenario is a realistic approach to the knotty problems of the world food crisis. The first element is the fundamental thesis of the climate mandate: There is no realistic prospect that world climates will become less variable. Thus with world population increasing steadily, there are bound to be severe regional crop failures that must be made up by food from the United States, Canada, and a few other food-surplus nations.

Our scenario calls for the creation of a world food emergency reserve for the inevitable local catastrophes. This reserve can be in money or in stored grain. Money is probably the more efficient way to store food-acquisition power, since cash is less perishable than grain. For the 25-year future, only the United States and Canada, in all probability, will be able to meet the demands for grain of all who can pay plus all who are in dire need. Thus emergency food-purchase power for hungry nations that cannot pay must be a key component of this strategy. Since the United States and Canada now have the major role in producing food over and beyond their domestic needs, the development of policies to guide the use of this food power will be fully as sensitive as those governing the OPEC role in oil. Furthermore, it is not at all certain where the money should come from to establish the reserve, although clearly it cannot be U.S. and Canadian charity. As with oil, producers and users need to be integrally involved in developing the governing policies for world food distribution in crisis and in view of the differences in the purchasing power of different world regions. Our scenario calls for joint U.S.-Canadian leadership in this matter, but it must be a far broader international effort.

The second element of our scenario deals with agricultural self-reliance for the developing nations. What strategies are best? Should the governing element be machine-intensive or labor-intensive? In this domain a serious clash of economic and political views between communist and capitalist nations seems inevitable. The communist nations regard the right to gainful employment as equal in importance to the rights to nutrition, education, health, and perhaps even superior to personal freedom or social dissent for individuals. The United States values political freedom and peaceable dissent as a *sine qua non*. However, neither the free-enterprise world, with its corporate multinational distributors of technology for development, nor the centrally controlled economies, with their pilot projects and advising teams in developing countries, have been fully successful in dealing with the world food crisis.

In our scenario, we envision a new cooperative approach to the problems of the Third World that calls for joint Soviet-U.S. teams of development advisors that would come in at the invitation of Third World governments to establish advisory mechanisms, pilot projects, demonstration farms, and similar efforts aimed at promoting overall self-reliance for participating nations. These teams might also include other national participants in addition to those from the Soviet Union and the

United States. Specifically, we anticipate as a start the creation of small teams of perhaps six Soviet and six U.S. advisors (and perhaps members from other nations as well) who will bring together their scientific, technological, social, and economic knowledge and experience. The team members would go together, by invitation, to specific developing countries or regional groups of countries to offer their consultation in the common belief that solution to the self-reliance issue in developing nations and regions will lessen world tension as well as promoting human welfare. Obviously, there will be points of agreement and points of difference in the advice and consultation offered. In many instances, the work of these advisory teams could ultimately lead to creation of institutes for training and demonstration that will help develop appropriate national or regional infrastructures to optimize the food producing and distributing capabilities of the developing nations.

A third element of our scenario involves a firm international commitment to filling the fundamental human needs for adequate nutrition in the face of the clear and recurring year-to-year and decade-to-decade variability of climate and thus of the food-producing capabilities of different regions. A realistic commitment will make difficult demands on food-surplus nations, primarily the United States and Canada. The people of these countries will have to sacrifice, to a degree, by paying higher prices for food in order to maintain adequate production to meet world needs for food in emergencies. But such a commitment will also make equally difficult foreign currency demands on other industrial nations, such as the Soviet Union, which are unlikely in the next decade or two to become self-sufficient in food production and will remain dependent on imports. The Third World nations with inadequate resources and climate-induced food shortages will also be called on for accommodation rather than simple confrontation with the food-rich nations. But this accommodation must not force the needy nation to abandon its goals of racial and economic self-reliance and become a victim of food imperialism.

Another important element of the scenario is critical analysis of the policy choices that the climate mandate imposes. Many careful studies of development policy fill the shelves of the world's libraries. All too often, however, their recommendations are not adopted, because they have neither involved participants from the region under consideration nor integrated the political, cultural, economic, and ethical issues with the scientific-technical alternatives. Far too often they do not offer realistic and useful options to the practical politicians who make national policies and strategies.

Some organizations are beginning to see the need for analyses of the whole world food problem in transdisciplinary studies that involve participants from the regions under consideration. An excellent example is a current study headed by the Argentine scientist Rolando Garcia, working at the Graduate Institute for Advanced International Studies in Geneva under the auspices of the International Federation of Institutes for Advanced Study. This study is examining the climatic and economic events of 1972, when food production turned down and food prices spiraled. Teams from Africa, Latin America, Southeast Asia, and other regions are working together to determine whether or not 1972 was really a disastrous year meteorologically or whether some modest and not unusual climatic fluctuations that year simply exceeded the elasticity of the existing combination of political, commercial, and cultural forces. They are trying to find lessons for the future in the 1972 experience.

There are many critically important issues that call for this kind of imaginative analysis involving participants from affected regions. One such study is needed to define the circumstances in which labor-intensive agriculture is preferable to machine-intensive agriculture; the answer is not so clear as one might think. Another would examine criteria for optimum land use in terms of climate, population, cultural patterns, firewood and fiber needs, urbanization, and, of course, food. Still another would consider potential incentives to improve the quality of rural life to reverse the headlong plunge to urbanization that characterizes most of the world. Our scenario envisions a number of such studies designed to identify political options and specify the costs and benefits of each alternative realistically for decision makers in the real world.

This brief scenario obviously is not a neat prescription to cure the chronic malady of hunger that afflicts a large part of the earth's population. We have not included some important efforts that are already underway—national and international research programs on plant genetics and climate forecasting, for example. Our principal point is that there is a critical need to consider the world food problem as a whole and to analyze it in terms of political, economic, and social options. This whole-problem approach has been badly neglected in the past. Piecemeal efforts, regardless of how efficiently each is conducted within the narrow context of its particular goals and methods, will not provide workable solutions.

The earth is marvelously bountiful. It should have the ability to produce adequate food for a doubled or tripled world population if necessary. But even the job of feeding those who are now alive will require us to apply our scientific and technical knowledge responsibly and with great care and foresight in order to protect and preserve the resources of the planet. Clearly, we must recognize the climate mandate, which dictates that the earth's bounty will rise and fall from time to time and place to place in response to climatic fluctuations. If we heed the climate mandate, and if we accept the fact that the earth's people are bound together by mutual needs and expectations that must transcend our rivalries and contests, humanity should be able not only to survive but to prevail over the hunger and starvation that have threatened so many people for so many centuries.

Bibliography

Alexander, Tom. 1974. Ominous Changes in the World's Weather. *Fortune,* February 1974:90.

American Meteorological Society Council. 1976. Policy Statement of the American Meteorological Society on Weather Forecasting. *Bulletin of the American Meteorological Society* 57:1460–1461.

Anderson, Alan, Jr. 1975. The Green Revolution Lives. *The New York Times Magazine,* April 27, 1975:15.

Babin, Gilles. 1975. Blizzard of 1975 in Western Canada. *Weatherwise,* April 1975:70–75.

Bandeen, William R., and Stephen P. Maran, eds. 1975. *Possible Relationships Between Solar Activity and Meteorological Phenomena.* Washington, DC: National Aeronautics and Space Administration.

Beman, Lewis. 1976. A New Case for That Old Ever-Normal Granary. *Fortune,* April 1976:97.

Borchert, John R. 1971. The Dust Bowl in the 1970s. *Annals of the Association of American Geographers* 61:1–22.

Brooks, C. E. P. 1949. *Climate Through the Ages,* 2nd rev. ed. New York: Dover.

Brooks, C. E. P. 1950. *Climate in Everyday Life.* London: Ernest Benn.

Brown, Lester R. 1975. *The Politics and Responsibility of the North American Breadbasket* (Worldwatch Paper 2). Washington, DC: Worldwatch Institute.

Brown, Lester R. 1976. *World Population Trends: Signs of Hope, Signs of Stress* (Worldwatch Paper 8). Washington, DC: Worldwatch Institute.

Brown, Lester R., with Erik P. Eckholm. 1974. *By Bread Alone.* New York: Praeger.

Bryson, Reid A. 1973. *Climatic Modification by Air Pollution, II: The Sahelian Effect* (Report 9, Institute for Environmental Studies). Madison: University of Wisconsin.

Bryson, Reid A. 1974a. A Perspective on Climatic Change. *Science* 184:753–759.

Bryson, Reid A. 1974b. *World Climate and World Food Systems, III: The Lessons of Climatic History* (Report 27, Institute for Environmental Studies). Madison: University of Wisconsin.

Bryson, Reid A. 1977. Quoted in interview in *The National Observer,* May 30, 1977.

Business Week. 1976. Weather Turns World Economics Topsy-Turvy. *Business Week,* August 2, 1976:48–50.

Calder, Nigel. 1975. *The Weather Machine.* New York: Viking.

Calder, Nigel. 1976. The Cause of Ice Ages. *New Scientist*, December 9, 1976:576–578.

Central Intelligence Agency. 1974. *Potential Implications of Trends in World Population, Food Production, and Climate*. Washington, DC: CIA.

Claiborne, Robert. 1970. *Climate, Man, and History*. New York: Norton.

CLIMAP. 1976. The Surface of the Ice-Age Earth. *Science* 191:1131–1144.

Conrat, Maisie, and Richard Conrat. 1977. *The American Farm*. Boston: Houghton Mifflin.

Donn, William L., and Maurice Ewing. 1966. A Theory of Ice Ages III. *Science* 152:1706–1712.

Dumond, Dwight Lowell. 1947. *America in Our Time*. New York: Holt.

Eckholm, Erik P. 1976. *Losing Ground*. New York: Norton.

Eddy, John A. 1975. A New Look at Solar-Terrestrial Relationships (Review paper presented at the 146th Meeting of the American Astronomical Society, San Diego, CA, August 18, 1975).

Eddy, John A. 1977. A Practical Question in Astronomy. *Science* 195:670–671.

Ehrlich, Paul R. 1968. *The Population Bomb*. New York: Ballantine.

Emery, F. V., and C. G. Smith. 1976. A Weather Record from Snowdonia, 1697–98. *Weather* 31:142–150.

Emiliani, Cesare. 1958. Ancient Temperatures. *Scientific American* 198, 2:67–78.

Glantz, Michael H. 1976a. *Value of a Reliable Long Range Climate Forecast for the Sahel: A Preliminary Assessment*. Boulder, CO: National Center for Atmospheric Research.

Glantz, Michael H. 1976b. *The Politics of Natural Disaster: The Case of the Sahelian Drought*. New York: Praeger.

Graff, John V., and Joseph H. Strub. 1975. The Great Upper Great Plains Blizzard of January 1975. *Weatherwise*, April 1975:66.

Grant, James P. 1977. Meeting the Basic Needs of the Poorest Billion—The World Can and Must Afford It. *United Nations Development Forum* 4, 7:1–2.

Greene, Wade. 1975. Triage. *The New York Times Magazine*, January 5, 1975:9.

Halacy, D. S., Jr. 1968. *The Weather Changers*. New York: Harper & Row.

Hammond, Allen L. 1974. Modeling the Climate: A New Sense of Urgency. *Science* 185:1145–1147.

Hammond, Allen L. 1976. Solar Variability: Is the Sun an Inconstant Star? *Science* 191:1159–1160.

Hardin, Garrett. 1968. The Tragedy of the Commons. *Science* 162:1243–1248.

Hardin, Garrett. 1974. Living on a Lifeboat. *Bioscience* 24:561–568.

Harrison, Paul. 1977. Beyond the Green Revolution. *New Scientist* June 9, 1977:575–578.

Hays, J. D., John Imbrie, and N. J. Shackleton. 1976. Variations in the Earth's Orbit: Pacemaker of the Ice Ages. *Science* 194:1121–1132.

Heady, Earl O. 1976. The Agriculture of the U.S. *Scientific American* 235, 3:106–127.

Henz, John F., and Vincent R. Sheetz. 1976. The Big Thompson Flood of 1976 in Colorado. *Weatherwise*, December 1976:278–285.

Hopper, W. David. 1976. The Development of Agriculture in Developing Countries. *Scientific American* 235, 3:196–205.

Institute of Ecology, The. 1976. *Impact of Climatic Fluctuation on Major North American Food Crops*. Washington, DC: The Institute of Ecology.

Interdepartmental Committee for Atmospheric Sciences. 1974. *Report of the Ad Hoc Panel on the Present Interglacial.* Washington, DC: National Science Foundation.

Jennings, Peter R. 1976. The Amplification of Agricultural Production. *Scientific American* 235, 3:180–194.

Kellogg, W. W., and S. H. Schneider. 1974. Climate Stabilization: For Better or Worse? *Science* 186:1163–1172.

Kennett, James P., and Robert C. Thunell. 1975. Global Increase in Quaternary Explosive Volcanism. *Science* 187:497–503.

LaMarche, Valmore. 1975. Climatic Clues from Tree Rings. *New Scientist*, April 3, 1975:8–11.

Lamb, H. H. 1966. *The Changing Climate.* London: Methuen.

Lamb, H. H. 1972. *Climate: Present, Past and Future.* London: Methuen.

Lamb, H. H. 1974. *The Current Trend of World Climate—A Report on the Early 1970's and a Perspective.* Norwich, England: University of East Anglia.

Landsberg, Helmut E. 1970. Man-Made Climatic Changes. *Science* 170:1265–1274.

Landsberg, Helmut E. 1976. Whence Global Climate: Hot or Cold? An Essay Review. *Bulletin of the American Meteorological Society* 57:441–443.

Lansford, Henry H. 1972. Weather Modification in the High Plains Region: Some Public-Policy Issues. In *The High Plains: Problems of Semiarid Environments* (Contribution No. 15 of the Committee on Desert and Arid Zones Research, Southwestern and Rocky Mountain Division, American Association for the Advancement of Science). Fort Collins: Colorado State University.

Lansford, Henry, rapporteur. 1975. The Policy Implications of Food and Climate Interactions (Summary of an IFIAS Project Workshop held at Aspen Berlin, February 5–7, 1975). Boulder, CO: Aspen Institute for Humanistic Studies.

Lansford, Henry, rapporteur. 1976. Climate Change, Food Production, and Interstate Conflict: A Bellagio Conference, 1975. New York: The Rockefeller Foundation.

Lappé, Frances Moore, and Joseph Collins. 1976. More Food Means More Hunger. *United Nations Development Forum* 4, 8:1–2.

Lappé, Frances Moore, and Joseph Collins, with Cary Fowler. 1977. *Food First.* Boston: Houghton Mifflin.

Le Roy Ladurie, Emmanuel. 1971. *Times of Feast, Times of Famine.* Garden City, NY: Doubleday.

Lenefsky, David, rapporteur. 1974. The Impact on Man of Climate Change (Report of an IFIAS Workshop held at the University of Bonn, May 6–10, 1974). Boulder, CO: Aspen Institute of Humanistic Studies.

Mason, B. J. 1977. Has the Weather Gone Mad? *The New Republic*, July 30, 1977:21–23.

Mayer, Allen J. 1976. A World Praying for Rain. *Newsweek*, July 19, 1976:66–67.

Mayer, Jean. 1976. The Dimensions of Human Hunger. *Scientific American* 235, 3:40–49.

Monin, Andrei. 1972. *Weather Forecasting as a Problem in Physics.* Cambridge: MIT Press.

Namias, Jerome. 1978. Long-Range Weather and Climate Predictions. In *Geophysical Predictions.* Washington, DC: National Academy of Sciences.

National Academy of Sciences. 1973. *Weather and Climate Modification.* Washington, DC: National Academy of Sciences.

National Academy of Sciences. 1975. *Understanding Climatic Change*. Washington, DC: National Academy of Sciences.

National Academy of Sciences. 1976. *Climate and Food*. Washington, DC: National Academy of Sciences.

National Oceanic and Atmospheric Administration. 1973. *The Influence of Weather and Climate on United States Grain Yields: Bumper Crops or Droughts*. Washington, DC: National Oceanic and Atmospheric Administration.

Newell, Reginald E. 1974. The Earth's Climatic History. *Technology Review*, December 1974:31–45.

Paddock, William, and Paul Paddock. 1967. *Famine—1975!* Boston: Little, Brown.

Pearson, Allen, and Frederick P. Ostby, Jr. 1975. The Tornado Season of 1974. *Weatherwise*, February 1975:5–7.

Ponte, Lowell. 1976. *The Cooling*. New York: Prentice-Hall.

President's Science Advisory Committee. 1967. *The World Food Problem*. Washington, DC: U.S. Government Printing Office.

Rothschild, Emma. 1977. Is It Time to End Food for Peace? *The New York Times Magazine*, March 13, 1977:15.

Schneider, Stephen H., and Clifford Mass. 1975. Volcanic Dust, Sunspots, and Temperature Trends. *Science* 190:741–746.

Schneider, Stephen H., with Lynne E. Mesirow. 1976. *The Genesis Strategy*. New York: Plenum.

Scrimshaw, Nevin S. 1974. The World-Wide Confrontation of Population and Food Supply. *Technology Review*, December 1974:13–19.

Silverberg, Robert. 1969. *The Challenge of Climate*. New York: Meredith.

Siscoe, George L. 1976. Solar-Terrestrial Relations: Stone Age to Space Age. *Technology Review*, January 1976:27–37.

Steele, Richard. 1977. The Deep Freeze! *Newsweek*, January 31, 1977:34–45.

Sugg, Arnold L., and Leonard G. Pardue. 1970. The Hurricane Season of 1969. *Weatherwise*, February 1970:13–15.

Thompson, Herbert J. 1969. The James River Flood of August 1969 in Virginia. *Weatherwise*, October 1969:180–183.

Thompson, Louis M. 1975. Weather Variability, Climatic Change, and Grain Production. *Science* 188:535–541.

Thompson, Philip D., and Robert O'Brien. 1965. *Weather*. New York: Time-Life Books.

Time. 1974. The World Food Crisis. *Time*, November 11, 1974:66–83.

Time. 1975. Poor vs. Rich: A New Global Conflict. *Time*, December 22, 1975:34–42.

Time. 1977. The Big Freeze. *Time*, January 31, 1977:22–28.

Trager, James. 1973. *Amber Waves of Grain*. New York: Arthur Fields.

United States Department of Agriculture. 1974. *The World Food Situation and Prospects to 1985* (Foreign Agricultural Economic Report No. 98). Washington, DC: U.S. Department of Agriculture.

U.S. News. 1974a. Formula for World Famine? *U.S. News & World Report*, January 28, 1974:50–52.

U.S. News. 1974b. Hot Debate: What U.S. Owes to the World's Hungry. *U.S. News & World Report*, December 16, 1974:20–22.

U.S. News. 1976. Europe's Drought Sears Crops, Farmers, Tourists. *U.S. News & World Report*, July 26, 1976:50–51.

U.S. News. 1977a. Winter's Toll: Worse to Come. *U.S. News & World Report*, February 14, 1977:19–20.

U.S. News. 1977b. The Spreading Impact of the Worst Drought in Decades. *U.S. News & World Report*, March 7, 1977:55–56.

Wade, Nicholas. 1974a. Sahelian Drought: No Victory for Western Aid. *Science* 185:234–237.

Wade, Nicholas. 1974b. Green Revolution (I): A Just Technology, Often Unjust in Use. *Science* 186:1093–1096.

Walker, Martin. 1974. Drought. *The New York Times Magazine*, June 9, 1974:11.

Wharton, Clifford, Jr. 1977. Feeding the World: Are Politicians the Missing Link? *RF Illustrated*, September 1977:1.

White, Gilbert F., and J. Eugene Haas. 1975. *Assessment of Research on Natural Hazards*. Cambridge: MIT Press.

Wilcox, H. A. 1975. *Hothouse Earth*. New York: Praeger.

Wilcox, John M. 1976. Solar Structure and Terrestrial Weather. *Science* 192:745–748.

Willett, Hurd C. 1976. The Sun as a Maker of Weather and Climate. *Technology Review*, January 1976:47–55.

Wilson, Thomas W., Jr. 1974a. *World Food: The Political Dimension*. Washington, DC: Aspen Institute for Humanistic Studies.

Wilson, Thomas W., Jr. 1974b. *World Population and a Global Emergency*. Washington, DC: Aspen Institute for Humanistic Studies.

World Meteorological Organization. 1975. *Proceedings of the WMO/IAMAP Symposium on Long-Term Climatic Fluctuations*. Geneva: World Meteorological Organization.

World Meteorological Organization. 1976. *Papers Presented at the Second WMO Scientific Conference on Weather Modification*. Geneva: World Meteorological Organization.

Wortman, Sterling. 1976. Food and Agriculture. *Scientific American* 235, 3:30–39.

Index